Electric Drives

Electric Drives

Marcel Jufer

ISTE

WILEY

First published 2010 in Great Britain and the United States by ISTE Ltd and John Wiley & Sons, Inc.

ISTE Ltd
27-37 St George's Road
London SW19 4EU
UK

www.iste.co.uk

John Wiley & Sons, Inc.
111 River Street
Hoboken, NJ 07030
USA

www.wiley.com

Library of Congress Cataloging-in-Publication Data

Jufer, Marcel, 1941-
Electric drives / Marcel Jufer.
p. cm.
Includes bibliographical references and index.
ISBN 978-1-84821-217-6
1. Electric driving. 2. Electric motors. I. Title.
TK4058.J84 2010
621.46--dc22

2010022200

British Library Cataloguing-in-Publication Data
A CIP record for this book is available from the British Library
ISBN 978-1-84821-217-6

Printed and bound in Great Britain by CPI Antony Rowe, Chippenham and Eastbourne.

Table of Contents

Chapter 1

Introduction – Electric Drive Components

1.1. Definition

An *electric drive* is a *system* providing an electromechanical conversion using an electric motor and including all the peripherals necessary for transmission, supply and control.

The concept of a system, in opposition to the motor only, is characteristic of the electric drive. The quality of this system is evaluated based on the weakest component.

The electric drive must be adapted to the application considered, using some of its components. The *driven body* (pump, machine tool, tracer, computer peripheral, robot, etc.), if it is not an integral part of the drive, determines its characteristics via the specifications which follow.

1.2. Electric drive components

The main components of an electric drive are as follows (Figure 1.1):

– the *transmission*, which enables us to adapt the movement (rotating-linear conversion, for example), the speed, the resolution or the dynamics of the system;

– the *electric motor*, defined by its external characteristics and by regulation;

– the *supply*, which enables us to adapt the primary electric source to the motor;

– the *command* and the *regulation* which assure the control of the system dynamic behavior (positioning, speed, current, torque, etc.);

– one or several *sensors* which supply the information necessary for control;

– the *safety* and *protection* elements such as thermal protection, over-speed protections, overload, etc.

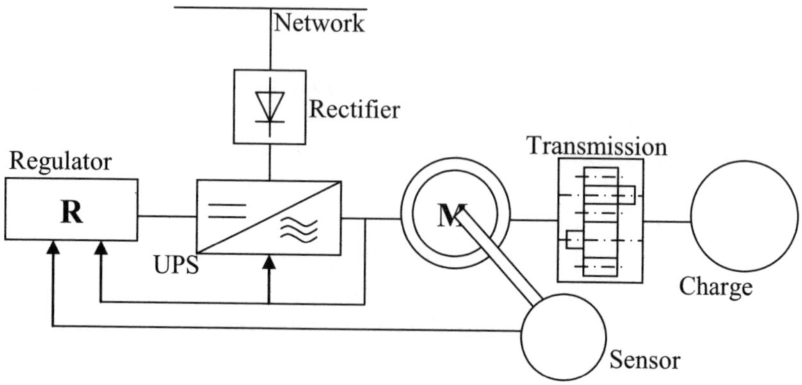

Figure 1.1. *Electric drive components*

In a traditional way, each of these elements is designed individually, to enable the widest possible range of applications. The current trend goes towards the integration of some of these components into the motor (transmission, supply, sensor), while increasing the flexibility of use by means of a command with a programmable microprocessor.

The approach of a system analysis of electric drive involves, first of all, a study of components:

– the driven body which enables us to define the specifications and the constraints compulsory for the electric drive;

– the transmission which enables the adaptation of the electric drive to the load;

– the motor, its supply, its command and its peripherals which are strictly connected and conditioned by the desired function.

There are two steps when choosing synthesis:

– the *weighting* of the diverse choice criteria. Certain factors are a priority in an evident way, while others are more difficult to control. In particular, the economic aspects are strongly connected to specific conditions such as quantities, manufacturing etc., thus with more *a priori* known difficulty;

– the process of *iteration* enables a comparison of several variants, so as to make a considered choice that is put into perspective.

The following chapters will cover the analytical aspects, while the final chapters put the focus on synthesis.

Chapter 2

Driven Bodies

2.1. Function of the driven body

Any driven body is characterized by a function: pumping, position transfer, machining, oscillating movement, speed control, etc. The *specifications* objective is to translate this function into the terms of electric drive. Furthermore, a certain number of *constraints*, bound to the environment associated with the driven body, can intervene: primary electric source, atmosphere and ambient temperature, dimensional constraints (diameter, length, mass), etc.

2.2. Reference or rated running

Most electric drives can be characterized by the *reference or rated running* of the load. This corresponds to the torque that the system can permanently supply, without overheating any of its components. It is related to a reference speed Ω_N. This reference speed is not the maximum possible speed. By rule and without particular precision, any drive has to withstand an over-speed of 20% of the rated speed.

$$\Omega_{max} = 1.2\ \Omega_N \tag{2.1}$$

Any rotating system is also characterized by a torque or a rated output with:

$$P_N = M_N \,.\, \Omega_N \tag{2.2}$$

As per usual, motors are defined by their torque rates and their speed for the small powers; and by their rated output and their speed for the higher powers (1 kW approximately).

For many driven bodies, the power (or the torque) and the rated speed are clearly defined values: a pump is generally working at constant or weakly variable power and speed. On the other hand, for other driven systems, this concept has no sense *a priori*; only the transient behavior is characterized and the concepts of rated power and speed are only defined in an equivalent way, *a posteriori*.

2.3. Transient behavior

Transient behavior is defined by an evolution of one or several parameters in a duration lower than or comparable to the biggest time constant of the system. Very frequently, this last one corresponds to the thermal time constant of the motor. Two types of transient behavior are to be considered:

– an exceptional behavior, occurring in a periodicity clearly greater than the thermal time constant. Typically, start up, braking with recovery or a short-term overload are cases of an exceptional transient regime. Generally, it does not influence the design of the motor, instead influencing the design of some peripherals: electric supply, specific command, protection, etc;

– a reference behavior consisting of a succession of periodic transitory regimes such as acceleration, transfer with constant speed, deceleration, dead time. In a given configuration – transmission, supply, control defined – equivalent rated design can be deduced. The rated speed corresponds to the maximum speed. The rated equivalent torque is defined as the one which leads to the same losses, in permanent regime, as the average losses in transient regime.

A value characterizes an intermittent use of an electric drive: it is the *rate of use* that establishes the ratio between the switch on time and the reference duration, as a rule 300 s. This concept is frequently used for actuators such as switches or pushers, less frequently for motors. Indeed, in this last case, the conditions and the relative duration of start up or braking play an important role, not only defined by the rate of use.

2.4. Specifications

2.4.1. *Basic data*

The specifications include as a rule the following basic information, enabling us to characterize the load in terms of electric drive:

– the rated or reference torque M_N;

– the rated speed Ω_N;

– the starting up torque M_d;

– the supply voltage and the type of primary supply source U_N (AC mains, DC mains, battery, etc.).

These elements constitute the basic data. A lot of additional information allows us to better characterize the system. We can classify them in *characteristics of regulation*, *start up*, *transitory* and relative to the *peripheral elements*.

2.4.2. *Characteristics of regulation*

These characteristics are relative to the motor behavior under load. They allow us to define ranges of speed, torque and resolution as well as the required precision.

In general, a domain of speeds is defined by two limit values, by a relative precision and by an acceptable maximum rate of oscillation:

– range of speed Ω_{min}-Ω_{max};

– relative precision p_r;

– oscillation rate Ω_{osc}/Ω_N.

Regarding the torque, a characteristic of maximum torque to be supplied according to the speed is generally imposed. The oscillatory torque due to reluctant effects or to the switching of the electronic supply is frequently limited:

– characteristic torque-speed under load $M(\Omega)$;

– maximum oscillatory torque M_{osc};

– or M_{osc}/M_N.

The resolution requirements are generally fixed to the stopping position, and are more rare in dynamic behavior. The resolution is generally fixed by increments (steps per revolution) or by an angle:

– Resolution α_p (angle) N_p (steps/rev).

– Precision (position difference relative to the step) $\Delta\alpha / \alpha_p$.

Some specifications also impose constraints in terms of regulation time constants.

2.4.3. *Start-up and braking characteristics*

The objective of the operation of start-up is to bring the motor into the range of functioning, in terms of speed and torque. This can implicitly be part of the usual domain of regulation, but can distinguish itself by constraints of specific resisting torque, acceleration torque, position transfer time or precision.

In terms of driven body, start-up is essentially characterized by the following parameters:

– the characteristic of resisting torque in the start-up mode $M_d\,(\Omega)$;

– the load inertia (or the mass for a linear movement) I, m_e;

– the frequency of the start-up (or the duration separating two start-ups), possibly the number of consecutive start-ups f_d;

– the maximum acceptable time of start-up $T_{d\,max}$.

Braking is sometimes the object of particular specifications, if it is not performed by slowing down in free mode. It is then characterized in a symmetric way with the start-up mode:

– characteristic of resisting torque in deceleration mode $M_f\,(\Omega)$;

– braking time (or characteristic speed-time) T_f;

– possibilities of energy saving $\Omega_f\,(t)$;

– stopping conditions: precision, resolution, oscillation, no overshooting.

2.4.4. *Transient characteristics*

Some systems are working in a mode corresponding to a succession of transient regimes: axes control of machine tools, robots, plotters, etc. They are generally characterized by a typical cycle. This will generally be defined as follows:

– characteristic of speed according to time $\Omega(t)$, $v(t)$;

– corresponding characteristic of load torque $M_r(t)$;

– cycle and timeout time or rate of use, T_c, T_m, τ_{ut}.

2.4.5. *Characteristics of the peripheral devices*

The specifications frequently impose certain constraints or characteristics relative to the peripheral devices.

Regarding the electrical supply, the conditions mainly concern the primary source of energy:

– nature of the primary source, type of source (AC or DC, network, battery, etc);

– voltage level U_N;

– limit current I_{max};

– rate of harmonic maximum rejection τ_h %;

– possibility of energy recovery or not.

The transmission can be *a priori* imposed only in principle, in execution (transmission ratio and inertia) or on the contrary left free. In case the transmission is imposed, the following elements have to characterize it:

– transmission ratio Ω_m, Ω_e Ω_m, x_e;

– equivalent transmission inertia or mass J_t, m_t;

– transmission efficiency $\eta_t(M, \Omega)$.

The sensors are sometimes directly integrated into the driven body. In that case, which is relatively rare, all characteristics of the sensor must be specified. In a more general case, the characteristics of the chosen sensor ensue mainly from information relative to the characteristics of regulation and from the requested resolution.

2.4.6. *Thermal aspect*

Any electric drive generates losses which must be evacuated by the environment. The characteristics and the possibilities of evacuation must be specified. The data necessary for this aspect are as follows:

– evacuation type of the losses:

- natural convection,
- forced convection,
- heat exchanger air-air,
- heat exchanger air-water,

– maximum ambient temperature T_{amb} (without precision, this value is 40°C);

– insulation class or maximum bearable temperature of the insulation (without precision, class B):

Y 90 [°C]

A 105

E 120

B 130

F 155

H 180

C 220

– particular corrosive, explosive atmospheres;

– degree of protection, corresponding in particular to the resistance to humidity and water.

2.4.7. *Constraints*

Unlike the characteristics, the constraints have a restrictive aspect with regard to the realization of specific components. They will generally be of a geometrical nature:

– dimensions constraints: diameter, length, volume, mass;

– constraints relative to a characteristic: inertia lower or greater than a limit value;

– supply constraints: limit current or power;

– thermal constraints: limited losses, extreme ambient temperatures;

– noise limits;

– electromagnetic compatibility limits;

– cost limits, etc.

2.4.8. *Specification list*

Table 2.1 below summarizes the characteristics of the driven body and its peripheral components. It offers a typical list which it is enough to fill and which allows us to fix the specifications and to define the criteria of choice and design.

1) Basic data	
1.1 Rated or reference torque	M_N
1.2 Rated speed	Ω_N
1.3 Start-up torque	M_d
1.4 Supply type and voltage	AC, DC; U_N

2) Characteristics of regulation	
2.1 Speed range	Ω_{min}
	Ω_{max}
2.2 Speed relative precision	p_r
2.3 Speed oscillation rate	Ω_{osc} / Ω_N
2.4 Torque – speed characteristic	$M\ (\Omega)$
2.5 Maximum oscillatory torque	M_{osc} / M_N
2.6 Resolution	N_p
2.7 Position accuracy	$\Delta\alpha / \alpha_p$
2.8 Time constants	
3) Characteristics of start-up and braking	
3.1 Start-up load torque	$M_d\ (\Omega)$
3.2 Workload inertia or mass	J_e
	m_e
3.3 Start-up frequency	f_d
3.4 Maximum start-up time	$T_{d^{max}}$
3.5 Braking load torque characteristic	$M_f\ (\Omega)$
3.6 Maximum braking time	$T_{f_{max}}$
3.7 Possibilities of energy saving by braking	
3.8 Stopping conditions: precision, resolution	
4) Transient characteristics	
4.1 Speed-time characteristic	$\Omega (t),\ v\ (t)$
4.2 Load torque-time characteristic	$M_{res}\ (t)$
4.3 Cycle time, rate of use	$T_c,\ \tau_{ut}$
5) Peripheral devices	
5.1 Nature of the primary source (AC, DC, battery, etc.)	
5.2 Voltage level	U_N
5.3 Maximum current	I_{max}
5.4 Rate of harmonic rejection	τ_h
5.5 Energy saving possibilities	

5.6 Transmission ratio	r	
5.7 Transmission inertia	J_t	
5.8 Transmission efficiency	$\eta\,(M,\Omega)$	
5.9 Sensors		
6) Thermal characteristics		
6.1 Cooling mode		
6.2 Ambient temperature	T_{amb}	
6.3 Insulation class ^{0}C	Class:	
	Y T_{max}	
	=	90°C
	A	105
	E	120
	B	130
	F	155
	H	180
	C	220
6.4 Particular atmosphere		
6.5 Degree of protection		
7) Constraints		
7.1 Dimension constraints		
7.2 Inertia		
7.3 Supply		
7.4 Thermal		
7.5 Noise		
7.6 EMC		
7.7 Others		
8) Economic aspects		
8.1 Production quantity		
8.2 Delivery time		
8.3 Costs		
9) Miscellaneous		

Table 2.1. *Summary of the characteristics of the driven body and its peripheral components*

Chapter 3

Transmission

3.1. Transmission types and characterization

3.1.1. *Rotating-rotating transmissions*

The conversion of a *rotating* movement in another rotating movement has several possible objectives:

– reduction or increase of speed;

– resolution increase;

– torque increase;

– change of rotation direction.

Such a transmission is mainly characterized by a transmission ratio r, defined as follows:

$$r = \Omega_m / \Omega_e \qquad [3.1]$$

where m is the motor index; and e is the driven body index.

$$r = \alpha_m / \alpha_e \qquad [3.2]$$

For an ideal system, without losses, we have:

$$r = M_e / M_m \quad [3.3]$$

For a real system, this last relation is corrected as follows:

$$M_e = M_m \, \eta_r = r \, M_m - \eta_r \quad [3.4]$$

with:

η_r = transmission efficiency;

M_r = equivalent friction torque of the transmission.

The expression above corresponds to a conversion motor – driven body. For an inverse conversion (braking), it becomes:

$$M_m = \frac{M_e}{r}\eta_r = \frac{M_e}{r} - M_r' \quad [3.5]$$

By preservation of the kinetic energy, we have the following expression for the inertia:

$$J'_m = r^2 \, J_m \quad [3.6]$$

with:

J'_m = motor equivalent inertia, reported to the driven body.

Figure 3.1. *Gears (helical, planetary, conical)*

The realization of a rotating-rotating transmission can appeal to very different technologies, as shown in Figures 3.1 and 3.2:

– gearings;

– pulleys and belt (trapeze shape, notched, etc.);

– pulleys and cable.

Figure 3.2. *Transmissions with pulleys and belt*

3.1.2. *Rotating-linear transmissions*

As for the rotating-rotating conversion, a rotating-*linear* conversion has various possible objectives, in addition to the change of movement:

– conversion and control of speed;

– conversion and control of position;

– force generation.

This transmission type can be characterized by a ratio of transmission k, defined as follows:

$$k = \Omega_m / v_e \text{ [rad/m]} \quad [3.7]$$

$$k = \alpha_m / \Delta x_e = 2\pi / step \quad [3.8]$$

α_m = angle moving

Δx_e = linear moving

step = linear step

Ideally, without losses:

$$k = F_e / Mm \quad [3.9]$$

For such a system, in motor mode:

$$F_e = kM_m\eta_k = kM_m - F_r \quad [3.10]$$

with:

F_r = friction force, equivalent to the transmission.

In braking mode:

$$M_m = \frac{F_e}{k}\eta_k = \frac{F_e}{k} - M'_r \quad [3.11]$$

The mass equivalent to the inertia of the motor is:

$$m'_m = k^2 J_m \quad [3.12]$$

A rotating-linear transmission can be realized using various techniques:

– screw, ball-screw, roller-screw;

Figure 3.3. *Ball- and roller-screw*

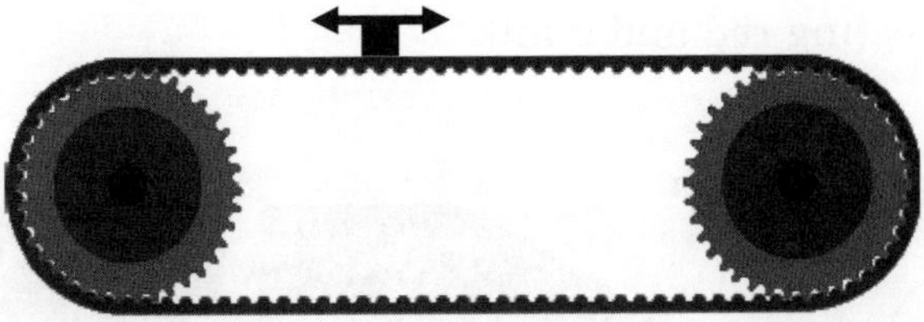

Figure 3.4. *Rotating-linear with pulleys and notch belt*

– capstan;

– gearing and rack;

Figure 3.5. *Gearing and rack*

– cam;

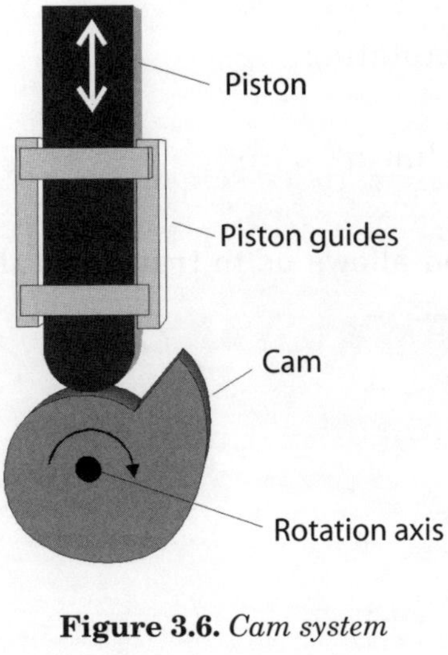

Figure 3.6. *Cam system*

– connecting rod and crank.

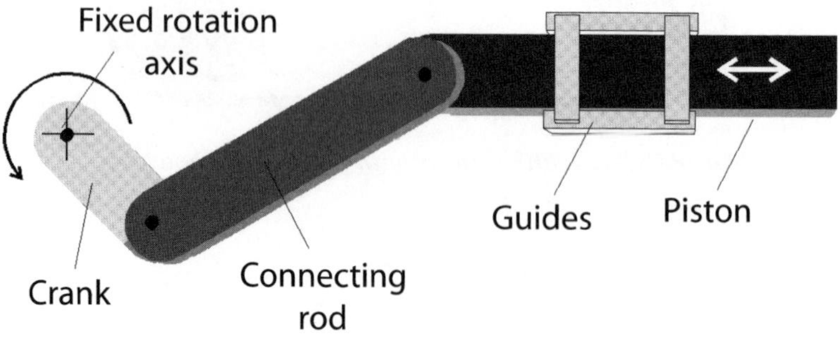

Figure 3.7. *Rod system*

3.2. Resolution

3.2.1. *Characteristics*

The resolution of a rotating or linear electric drive, is defined by the angular elementary step α_p, linear x_p or by the number of elementary steps on an angular distance (2π) or linear distance given by:

$$N_p = 2\pi / \alpha_p \text{ rotating} \qquad [3.13]$$

$$N_p = x_{ref} / x_p \text{ linear} \qquad [3.14]$$

A transmission allows us to transform the resolution. For a rotating system:

$$\alpha_{pe} = \alpha_{pm} / r \qquad [3.15]$$

$$N_{pe} = N_{pm} \,.\, r \qquad [3.16]$$

For a linear system:

$$x_{pe} = \alpha_{pm} / k \qquad [3.17]$$

If $x_{ref} = 2\pi / k$:

$$N_{pe} = N_{pm} \qquad [3.18]$$

The precision is defined as the maximum position gap, reported in the step:

$$p = \Delta\alpha_p / \alpha_p \qquad [3.19]$$

3.2.2. *Choice criteria*

For a motor, the *resolution* can be intrinsic or defined by an associated sensor. It is intrinsic for a step motor or a self-commutated synchronous motor:

– for a step motor, the usual resolution N_p is included between 2 and 500 steps/rev;

– for a self-commutated synchronous motor, the implicit resolution is defined by the following relations:

$$N_p = 2\ m\ p \text{ commutation at } 120^\circ \text{ or at } 180^\circ \qquad [3.20]$$

$$N_p = 4\ m\ p \text{ commutation at } 150^\circ \qquad [3.21]$$

A *sensor* allows us to obtain practically any resolution, until an order of 100,000 increments per revolution. However, this range can be broken down into two domains:

– resolution ≤ 1,000 increments per revolution, which can be considered as inexpensive for sensors produced in large series and integrated;

– resolutions > 1,000, which is expensive.

Other solutions are possible, such as indirect sensors bound to the back-EMF voltage created by the movement [JUF 92, 95a, 95b].

3.2.3. *Limits*

In theory, the use of a transmission allows us to obtain any value of resolution. In practice, several physical limits restrict these possibilities:

– the play introduced by any transmission. The make-up of systems of play allow us to reduce this effect, but by increasing the friction and the wear;

– the elasticity of any system of connection and transmission, which introduces an imprecision associated with the load variations.

3.3. Speed adaptation

An *adaptation of speed* can be imposed by the necessity of reaching extreme, very high speeds or on the contrary very low speeds. However, more generally, the adaptation of speed by a transmission can be simply imposed by economic criteria. Indeed, as seen in Chapter 4, the volume and the weight of a motor, thus also its cost, are inversely proportional to its nominal speed, for a given power.

$$V_m \sim 1 / \Omega_N \qquad [3.22]$$

It is thus often more interesting to use a fast motor and a transmission to reduce the speed, rather than a slow motor and a direct transmission. In addition, most of the motors produced by series are characterized by a nominal speed included between 600 and 3,000-6,000 rpm. For a very different speed range, an adaptation or a special execution is necessary.

3.4. Dynamic behavior

3.4.1. *Aim*

The objective of the study of *dynamic behavior* is to bring the role of the transmission into the global drive behavior, in particular in acceleration mode. The choice of transmission ratio influences the acceleration and braking times. A specific optimization can thus be realized.

3.4.2. *Dynamic equations*

The main job of numerous systems is to assure performances of acceleration and deceleration, practically without constant speed. This is a common case for positioning systems.

Generally, the movement equation of a rotating-rotating system can be written as:

$$(J_e + r^2 J_m)\frac{d\Omega_e}{dt} = r\, M_m - M \qquad [3.23]$$

where:

M_m = motor torque;

M_e = load torque (live and friction torque).

A linear characteristic for the motor torque according to the speed is considered in Figure 3.8.

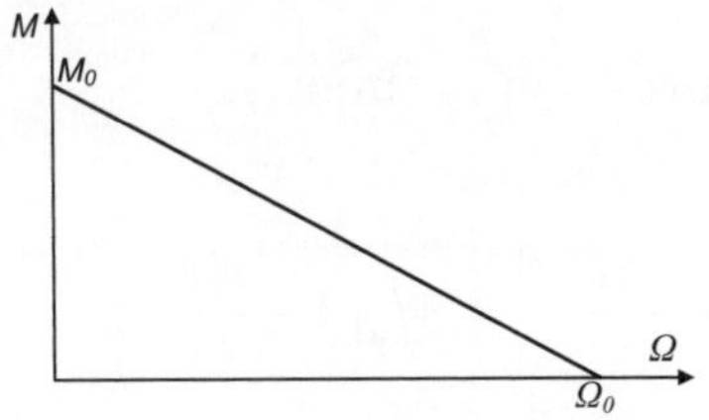

Figure 3.8. *Linear torque-speed characteristic*

This is the natural characteristic of a DC motor with separate excitation. For many other motors, the first part of the torque characteristic can also be approached by a straight line. For example, Figure 3.9 corresponds to a small induction motor while Figure 3.10 corresponds to a large-sized induction motor. Figure 3.13 illustrates the case of a brushless DC motor with current limitation.

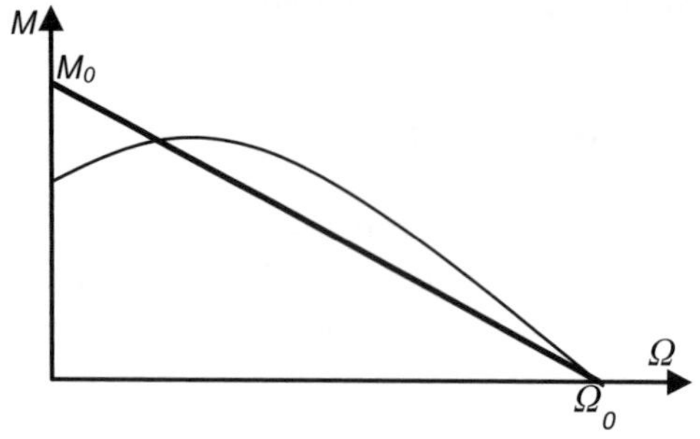

Figure 3.9. *Linear approximation (small induction motor)*

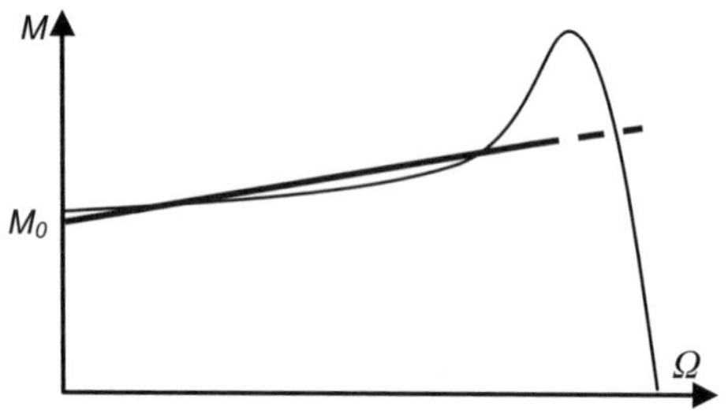

Figure 3.10. *Linear approximation (large induction motor)*

The torque expression is thus:

$$M_m = M_0\left(1 - \frac{\Omega_m}{\Omega_0}\right) = M_0\left(1 - \frac{r\Omega_e}{\Omega_0}\right) \qquad [3.24]$$

The load torque can frequently be reduced to two terms:

M_s = constant load torque due to a dry friction or to a constant load (rotating-linear system, F_s);

$M_v = \xi \Omega_e$ = viscous friction torque (rotating-linear system: $F_v = \chi v_e$).

The movement equation becomes:

$$(J_e + r^2 J_m)\,\frac{d\Omega_e}{dt} = r\,M_o\,(1 - r\,\frac{\Omega_e}{\Omega_0}) - M_s - \xi\,\Omega_e \qquad [3.25]$$

$$(J_e + r^2 J_m)\,\frac{d\Omega_e}{dt} + \Omega_e\,(\frac{r^2 M_0}{\Omega_0} + \xi) = r\,M_0 - M_s \qquad [3.26]$$

The solution of this linear equation is:

$$\Omega = \Omega_{es}\left(1 - e^{-t/\mathrm{T}mec}\right) \qquad [3.27]$$

Figure 3.11 illustrates this expression by defining several characteristic variables as shown in the following:

– The *stabilized speed* (particular solution):

- for a rotating system:

$$\Omega_{es} = \frac{rM_0 - M_s}{\dfrac{r^2 M_0}{\Omega_0} + \xi} \qquad [3.28]$$

- for a linear system:

$$v_{es} = \frac{kM_0 - F_s}{\dfrac{k^2 M_0}{\Omega_0} + \chi} \qquad [3.29]$$

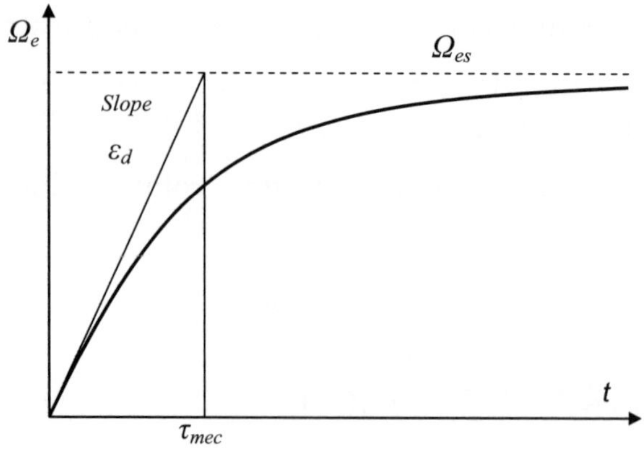

Figure 3.11. *Start up speed characteristic*

– The *mechanical time constant*

- for a rotating system:

$$\tau_{mec} = \frac{J_e + r^2 J_m}{\dfrac{r^2 M_0}{\Omega_0} + \xi} \qquad [3.30]$$

- for a linear system:

$$\tau_{mec} = \frac{m_e + k^2 J_m}{\dfrac{k^2 M_0}{\Omega_0} + \chi} \qquad [3.31]$$

– The *starting acceleration*.

- for a rotating system:

$\varepsilon_d = d\Omega/dt$, at $t = 0$

$$\varepsilon_d = \frac{rM_0 - M_s}{J_e + r^2 J_m} \qquad [3.32]$$

- for a linear system:

a_d = dv/dt, at t = 0

$$a_d = \frac{kM_0 - F_s}{m_e + k^2 J_m} \quad [3.33]$$

Figure 3.4 defines and represents these different characteristic values.

3.4.3. *Mechanical time constant*

The *mechanical time constant,* according to the expression above, makes sense only for the case of Figures 3.8 and 3.9. In a more general case, another definition must be given. The reference mechanical time constant can then be defined as follows:

$$\tau_{mec} = \frac{J_{eq}\Omega_N}{M_N} \quad [3.34]$$

with:

Ω_N = rating speed;

M_N = rating torque;

J_{eq} = equivalent inertia.

This time constant corresponds to the start-up time to reach the nominal speed by applying the rating torque to the system.

The mechanical time constant associated with the torque characteristic of Figure 3.8 presents an evolution according to the reduction ratio dependent on inertias and torques. This is represented in Figure 3.12.

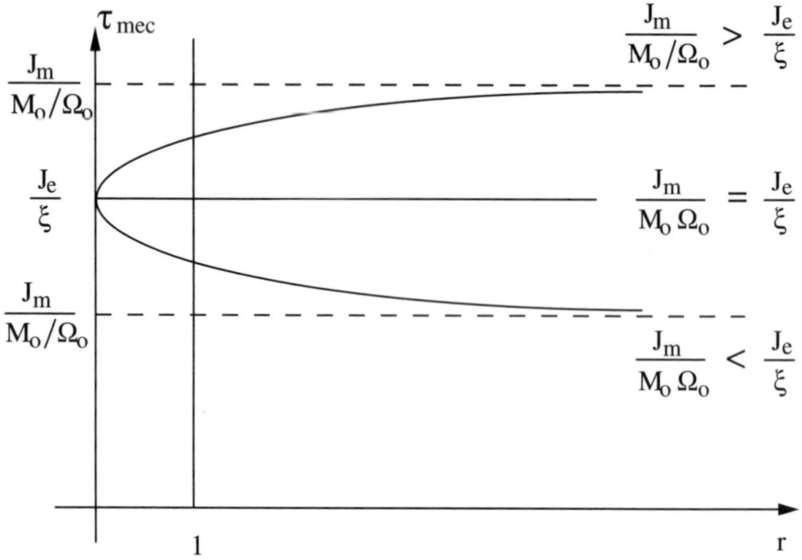

Figure 3.12. *Evolution of the mechanical time constant according to the reduction ratio*

Three cases are to be considered:

$$\frac{J_m}{M_0 / \Omega_0} > \frac{J_e}{\xi} \qquad [3.35]$$

The gearing has to be a speed multiplier. This then leads to a limit value equal to J_e / ξ.

$$\frac{J_m}{M_0 / \Omega_0} = \frac{J_e}{\xi} \qquad [3.36]$$

Gearing has no effect on the mechanical time constant.

$$\frac{J_m}{M_0 / \Omega_0} < \frac{J_e}{\xi} \qquad [3.37]$$

Gearing ensures a decrease of the mechanical time constant with an asymptotic value equal to:

$$\tau_{mec} = \frac{J_m}{M_0 / \Omega_0} \qquad [3.38]$$

This notion of the mechanical time constant is characteristic in start-up until the nominal value of the speed. To concretize the start-up performances leading to a speed which is only a fraction of the nominal speed, we shall resort to the acceleration.

3.4.4. *Acceleration*

The *acceleration* is characterized by the relations in section 3.4.2 ([3.30] and [3.31]). This is significant for systems of position transfer. This expression presents a maximum:

$$\frac{d\varepsilon_d}{dt} = \frac{M_0\left(J_e + r^2 J_m\right) - 2rJ_m\left(rM_0 - M_s\right)}{\left(J_e + r^2 J_m\right)^2} = 0 \qquad [3.39]$$

$$r^2 J_m M_o + 2\, r\, J_m\, M_s + M_o J_e = 0 \qquad [3.40]$$

A single solution makes sense: $r > 0$

$$r_0 = \frac{M_s}{M_0} + \sqrt{\left(\frac{M_s}{M_0}\right)^2 + \frac{J_e}{J_m}} \qquad [3.41]$$

For a rotating-linear system:

$$k_0 = \frac{F_s}{M_0} + \sqrt{\left(\frac{F_s}{M_0}\right)^2 + \frac{m_e}{J_m}} \qquad [3.42]$$

If the dry friction torque is not important, this expression becomes:

$$r_0 = \sqrt{\frac{J_e}{J_m}} \quad k_0 = \sqrt{\frac{m_e}{J_m}} \qquad [3.43]$$

This last expression leads us to choose a reduction ratio: such as, the motor inertia reported to the load is equal to that of the load.

$$r_0^2 J_m = J_e \qquad [3.44]$$

Example

A DC motor (Maxon 2128.910) is characterized by the following main values:

Ω_0= 5,200 rpm

J_m= 8.57 × 10^{-7} kgm^2

M_0= 1.8 × 10^{-2} Nm

This motor is intended for the drive of a daisy wheel (printer) presenting an inertia of 4.5 × 10^{-6} kgm^2. Dry and viscous frictions are not important. On average, such a printing head has to ensure a rotation of $\pi/4$. This angle can be broken down into a phase of acceleration on an angle of $\pi/8$ and a phase of deceleration on the second half of the angle ($\pi/8$).

Such a motor works practically only on the regimes of acceleration and deceleration.

The optimal acceleration is obtained for a gearing presenting a ratio equal to [3.42]:

$$r = \sqrt{\frac{J_e}{J_m}} = \sqrt{\frac{4.5.10^{-6}}{8.5710^{-7}}} = 2.29$$

In these conditions, the following values are obtained.

Gearing	With	Without
Acceleration ε_d [rad/s^2]	4.58×10^3	3.36×10^3
Acceleration time (π/8) [ms]	13	15.3
Peak speed Ω_e [rad/s]	59.5	51.4

3.5. Oscillatory torque

3.5.1. *Aim*

When the load is characterized by a torque varying periodically, several problems can result:

– the oscillation of the resultant speed must be reduced. This can be achieved by increasing the system inertia;

– as far as the instantaneous power presents important peak values of short duration (punching machines for example), an essential part of the corresponding energy can be removed from the kinetic energy stored in the rotating masses. The motor then works in a more or less continuous way on the renewal of the kinetic energy.

The principle of the cases above will be analyzed allowing us to choose the main parameters: motor inertia and power.

3.5.2. *Smoothing the speed oscillations*

A load characterized by an oscillating torque corresponds to the following torque expression, keeping only the fundamental term:

$$M_r = M_0 + \hat{M}_p \sin \omega_p t \qquad [3.45]$$

with:

M_0 = average resisting torque;

$\hat{M}_p$ = oscillating torque amplitude;

ω_p = oscillating torque.

The average motor torque M_m will be equal and opposite to the average resisting torque.

$$J \frac{d\Omega}{dt} = M_m - M_r = -\hat{M}_p \sin \omega_p t \qquad [3.46]$$

$$\Omega = \bar{\Omega} + \Delta\Omega \qquad [3.47]$$

with:

$\Delta\Omega$ = speed oscillation

$$\Delta\Omega = \frac{\hat{M}_p}{J\omega_p} \cos \omega_p t \qquad [3.48]$$

The expression of the speed oscillation amplitude is thus:

$$\Delta\hat{\Omega} = \frac{\hat{M}_p}{J\omega_p} \qquad [3.49]$$

If this maximum oscillation is imposed, the total inertia can be determined from:

$$J_t = \frac{\hat{M}_p}{\Delta\hat{\Omega}\,\omega_p} \qquad [3.50]$$

This fixes the necessary inertia for speed smoothing. If this value is higher than the load and motor inertia, the addition of a flywheel is indispensable.

3.5.3. *Torque pulses*

In the case of an electric drive associated with a driven body presenting important torque peaks of brief duration, two parameters must be determined in order to minimize the drive cost:

– a *flywheel* intended to store the kinetic energy;

– the motor power.

If the energy supplied in the driven body is taken from the kinetic energy, the motor does not have to be at constant speed:

– induction motor;

– DC motor;

– BLDC motor.

For the first two cases, the torque-speed characteristic is linear. For the last case, it can be approached by a straight line in the useful zone.

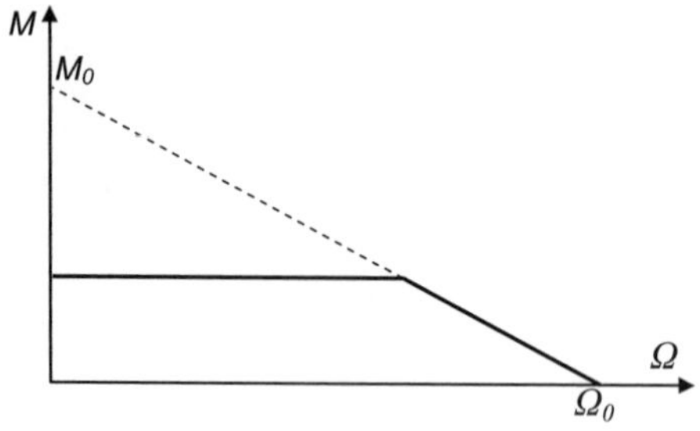

Figure 3.13. *DC and BLDC motor characteristics with current control*

The energy balance can be characterized as follows:

$$\Delta W_{\text{mec}} = \Delta W_{\text{kin}} \qquad [3.51]$$

ΔW_{mec} = useful mechanical energy per cycle

ΔW_{kin} = variation of kinetic energy by cycle

$$\Delta W_{kin} = \frac{1}{2} J_t \left(\Omega_1^2 - \Omega_2^2 \right) \qquad [3.52]$$

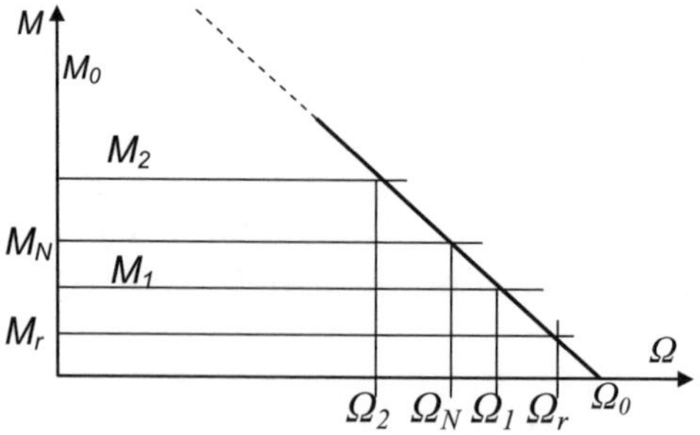

Figure 3.14. *Speed oscillation around the rating point*

with:

Ω_2 = maximum speed reached by the motor on a cycle;

Ω_1 = minimum speed reached by the motor on a cycle.

The deceleration phase is generally of brief duration compared with the period. The behavior is mainly governed by the load.

The acceleration phase corresponds to the following expression:

$$J\frac{d\Omega}{dt} = M_0(1-\Omega/\Omega_0)-M_r \quad [3.53]$$

The solution of this equation is:

$$\Omega(t) = \Omega_0(1-M_r/M_0)(1-e^{-t/\mathrm{T}mec})+\Omega_2 e^{-t/\mathrm{T}mec} \quad [3.54]$$

$$\Omega_1 = \Omega(T) \quad [3.55]$$

with:

T = period

$$\Omega(T) = \Omega_0(1-M_r/M_0)(1-e^{-T/\mathrm{T}mec})+\Omega_1 e^{-T/\mathrm{T}mec} = \Omega_1 \quad [3.56]$$

The following approach can be adopted with the aim of the parameter choice:

– choice of the motor power based on the energy balance:

$$P = \frac{\Delta W_{mec}}{T} + M_r(\Omega_1+\Omega_2)/2 \quad [3.57]$$

– choice of the speed Ω_1, such that the acceleration torque is not too low at the end of the period ($\Omega_1 > \Omega_r$), for example:

$$\Omega_1 \simeq \Omega_o - 2\,(\Omega_o - \Omega_r) = 2\,\Omega_r - \Omega_o \qquad [3.58]$$

Assuming:

$$\Omega_o - \Omega_1 \simeq 2\,(\Omega_o - \Omega_r) \qquad [3.59]$$

Then:

$$\Omega_o - \Omega_1 \simeq (\Omega_o - \Omega_N)/4 \qquad [3.60]$$

– choice of the speed Ω_2 is lower than the speed Ω_N ($M_2 > M_N$).

As a first estimate, we can write:

$$\left(M_1 + M_2\right)/2 = M_N \qquad [3.61]$$

$$\begin{aligned} M_2 &= 2M_N - M_1 \\ \Omega_2 &= 2\Omega_N - \Omega_1 \end{aligned} \qquad [3.62]$$

– the total inertia is determined by the expression of the kinetic energy:

$$J_t = \frac{2W_{cin}}{\left(\Omega_1^2 - \Omega_2^2\right)} \qquad [3.63]$$

By using the expression of the speed according to time, the choice of the values Ω_1, Ω_2, J_t must be verified, by checking the balance sheet of the motor losses. Using an iterative process, the final values can be determined.

3.6. Position transfer

3.6.1. *Aim*

A motor associated with a driven body, including a transmission (or not), has to assure a *position transfer* (angle

α_{tot}) practically at no load at a given time T. This operation is repetitive with a period T_{tot} with:

$$T_{tot} \geq T$$

This can thus imply a time out between two periods of duration:

$$T_{out} = T_{tot} - T \qquad [3.64]$$

The objective consists of choosing a profile such that the energy balance (copper losses) is the best possible to limit the heating and make the best use of the motor.

3.6.2. *Speed profile*

The simplest process consists of realizing three consecutive phases (Figure 3.15):

– an acceleration phase with a time t_1;

– an intermediate phase at constant speed with a time $T - t_1 - t_2$;

– a deceleration phase with a time t_2;

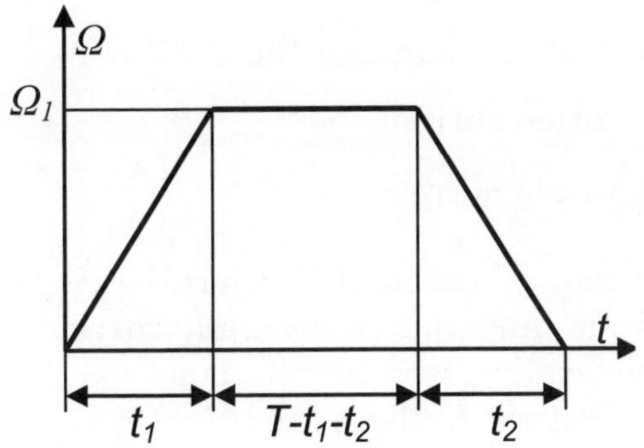

Figure 3.15. *Trapezoidal speed profile*

Ultimately the phase with constant speed can present a zero duration (Figure 3.16).

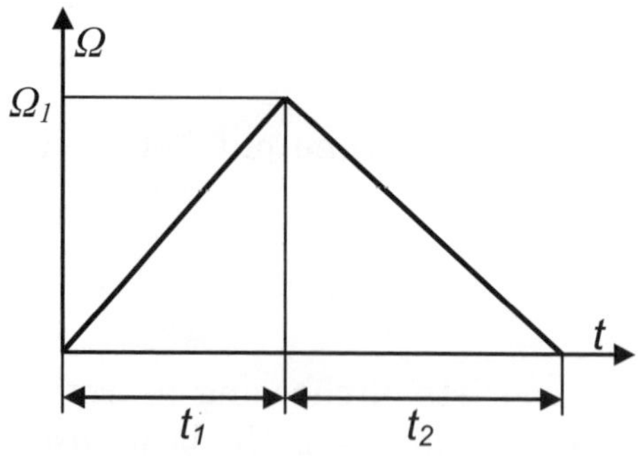

Figure 3.16. *Triangular speed profile*

The simplest way to control the acceleration (or the deceleration) consists of imposing a constant current. For a DC motor, with or without a collector, the torque is proportional to the current. In these conditions, the energy balance can be written as follows:

$$W_{Copper} = RI_1^2 t_1 + RI_2^2 t_2 \qquad [3.65]$$

where:

R = equivalent resistance at the DC level;

I_1 = acceleration current;

I_2 = deceleration current.

The condition of constant acceleration is obtained by a constant torque, and thus a constant current; the constant current is generally given as the same for the acceleration and the deceleration. In these conditions we can write:

$$W_{Copper} = 2RI_1^2 t_1 \qquad [3.66]$$

If the rotation angle α_{tot} is imposed, we have, using Newton's equations, at constant angular acceleration ε, with α_1 and α_2 the respective angles of acceleration and deceleration:

$$\alpha_1 = \frac{1}{2}\varepsilon t_1^2 = \alpha_2 \quad [3.67]$$

The maximum speed Ω_1 is:

$$\Omega_1 = \varepsilon t_1 \quad [3.68]$$

$$\alpha_{tot} = 2\alpha_1 + \Omega_1 (T - 2t_1) \quad [3.69]$$

$$\alpha_{tot} = \varepsilon . t_1 (T - t_1) \quad [3.70]$$

The variable ε is the angular acceleration, the expression of which becomes:

$$\varepsilon = \frac{\alpha_{tot}}{t_1 (T - t_1)} = \frac{M}{J} \quad [3.71]$$

with:

M = acceleration torque = $k\, I$;

J = total inertia;

k = torque constant [Nm/A].

$$I = \frac{M}{k} = \frac{J\alpha_{tot}}{k t_1 (T - t_1)} \quad [3.72]$$

The lost energy thus becomes:

$$W_{Copper} = 2R \frac{J^2 \alpha_{tot}^2}{k^2 t_1 (T - t_1)^2} \quad [3.73]$$

The system presents a minimum energy consumption for the following conditions:

$$\frac{dW_{Copper}}{dt_1} = 0 \quad [3.74]$$

The numerator of the expression has to be zero, thus:

$$3t_1^2 - 4t_1 T + T^2 = 0$$

The solutions are:

$$t_1 = T/3 \text{ and } t_1 = T \quad [3.75]$$

Only the solution $t_1 = T/3$ makes sense.

In these conditions, the energy dissipated in the equivalent resistance is then:

$$W_{J\min} = \frac{27RJ^2\alpha_{tot}^2}{2k^2T^3} \text{ with } t_1 = T/3 \quad [3.76]$$

As a comparison, for a triangle shape behavior (Figure 3.9), the result is:

$$W_{J\min} = \frac{16RJ^2\alpha_{tot}^2}{k^2T^3} = \frac{32}{27}W_{J\min} = 1.19W_{J\min} \quad [3.77]$$

with:

$t_1 = T/2$.

Chapter 4

Motors

4.1. Characterization

Since the development of the first electric motor, the DC current type with a collector, numerous motor variants have been conceived and applied in a relatively extensive way.

Two important motor types may be distinguished:

– the induction or asynchronous motors, for which the stator rotating magnetic field has a different speed to the rotor speed;

– the synchronous or auto-synchronous motors, for which the stator and rotor rotating fields both have the same speed. With the DC current motor, due to the role of the collector, it is the rotating field of the rotor that is synchronous with the stator, and thus fixed in the space.

Other characterization criteria can be considered:

– the generated movement, rotating or linear;

– the motor supply in opened or closed circuit.

4.2. Rotating and linear motors

The generation of a linear movement can be achieved by a rotating motor associated with a rotating-linear conversion (see section 3.1.2) or directly by a *linear motor,* generating a force and a translatory movement. Figures 4.1a and b show examples of this type of motor. Unlike a rotary motor, the concept of transmission, including an adaptation of speed or force, is not possible. Furthermore, a part of the motor (generally the mobile part), is directly integrated into the driven body.

The main applications are in the field of high speed transportation – in particular the systems with magnetic levitation called Maglev (Figure 4.1a) – machine tools (Figure 4.1b) and of numerous special applications.

a) b)

Figure 4.1. *a) Maglev vehicle with linear motor propulsion – MLX Japan; b) linear motor for machine-tools*

4.3. Induction motors

4.3.1. *Structure*

Most of the *induction or asynchronous motors* [JUF 95a] are made up of a *stator* with a polyphase winding (three or two phases) and a *rotor* formed by a non-isolated conductive cage, inserted into a stack of insulated laminations (Figure 4.2). Only the stator is supplied by an electric source, the rotor being indirectly supplied by a current resulting from the induction phenomenon.

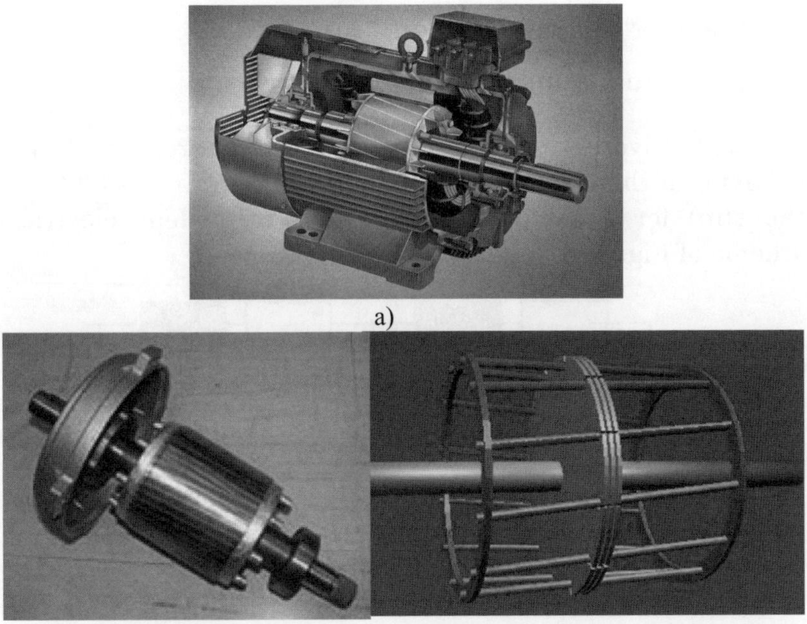

a)

b) c)

Figure 4.2. *a) Motor structure with stator, rotor and housing with blades; b) rotor of an asynchronous motor with aluminum injected cage; c) schematic representation of the cage and the rings*

4.3.2. *Equivalent electric scheme*

The characteristic elements of the induction motor are:

– the *slip* s with:

$$s = \frac{\Omega_s - \Omega}{\Omega_s}$$ = relative speed difference between stator rotating field and rotor;

– stator phase resistance R_s

– rotor equivalent phase resistance R'_r;

– stator leakage reactance $X_{\sigma s}$;

– the rotor equivalent leakage reactance $X'_{\sigma r}$;

– the mutual equivalent reactance X_h.

These diverse elements enable us to represent the behavior of the induction motor according to the speed (or of the slip) for a motor phase by the equivalent electrical scheme of Figure 4.3.

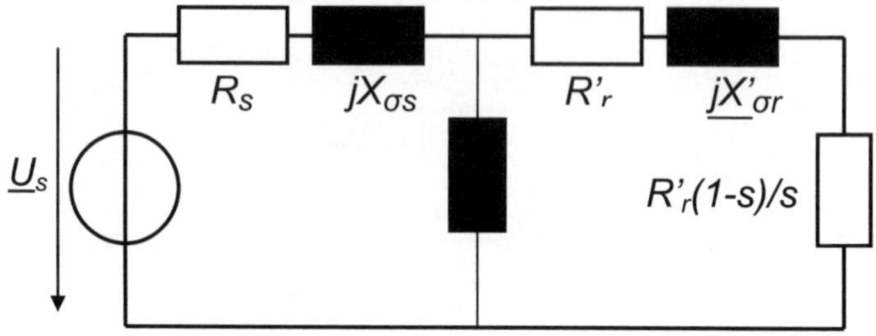

Figure 4.3. *Equivalent electrical scheme for an induction motor phase*

4.3.3. *Characteristics of current and torque with constant frequency*

The characteristics of the elements of an asynchronous motor are:

– the slip s;

– the characteristic of current according to the speed (or to the slip) with two characteristic points (Figure 4.4):

- the starting current I_d,

- the rating current I_N,

- the no-load current (zero torque) at synchronous speed I_0;

– the torque characteristic according to the speed or the slip, given by the following expression with three characteristic points (Figure 4.5):

- torque expression $M = \frac{3R'_r I'^2_r}{s\Omega_s}$, [4.2]

- the starting torque M_d,

– the maximum torque M_K corresponding to the slip s_K with:

$$s_K = \pm \frac{R'_r}{X_{cc}},$$ [4.3]

– rating torque M_N.

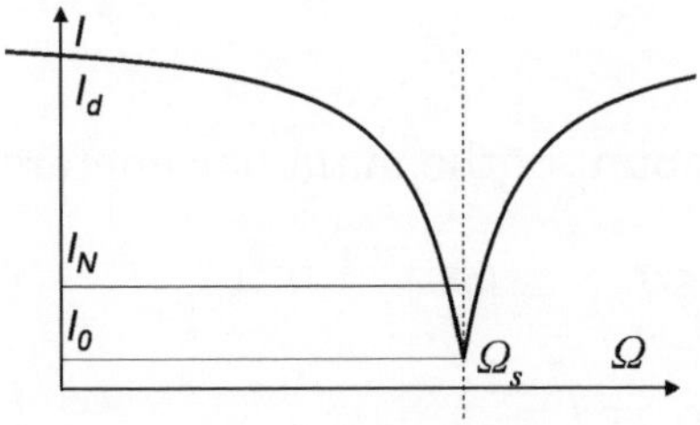

Figure 4.4. *Current characteristic according to the speed with constant voltage*

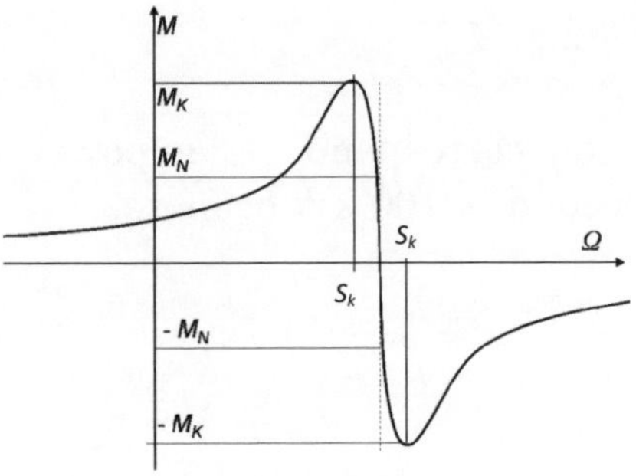

Figure 4.5. *Torque characteristic according to the speed with constant voltage*

For high slips ($|s| \geq 0.5$), the torque characteristic corresponds to a hyperbola with $M.s$ = constant (Figure 4.6 – 1).

For low slips, this characteristic becomes a straight line with $M = a.s$ (Figure 4.6 –2).

Between these two functions, the torque is assumed to be by a maximum (minimum) ± M_K for a slip ± s_K.

A good estimate of the torque is given by the following expression:

$$\frac{M}{M_K} = \frac{2}{\frac{s}{s_k} + \frac{s_K}{s}} \qquad [4.4]$$

The usual amount of the main parameters are:

$$I_d = 2.5 \div 6.I_N$$

$$I_0 = 0.5 \div 0.25.I_N \qquad [4.5]$$

$$M_d = 0.3 \div 1.M_N$$

$$M_K = 1.7 \div 2.5.M_N$$

The first value corresponds to a power of the order of 1 kW and the second to 100 kW or more.

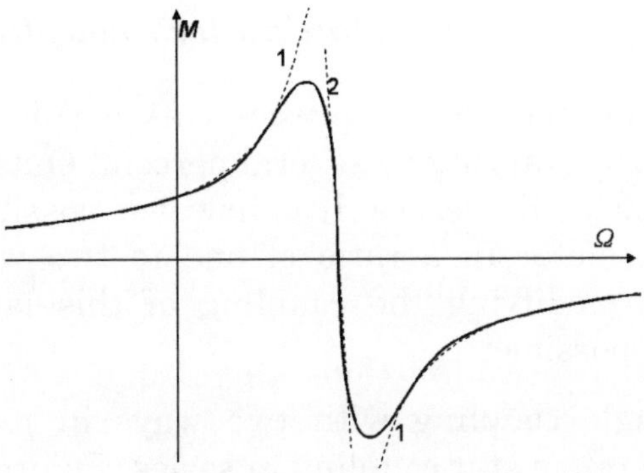

Figure 4.6. *Torque characteristic according to the speed: **1** – hyperbola zone; **2** – linear zone*

It is possible to show that for such a motor the rotor efficiency when iron losses are neglected, is worth [JUF 95a]:

$$\eta_r = \frac{P_{mec}}{P_{mec} + P_{Jr}} = 1 - s \qquad [4.6]$$

The overall efficiency of the motor η_m is thus lower than this value (Figure 4.7).

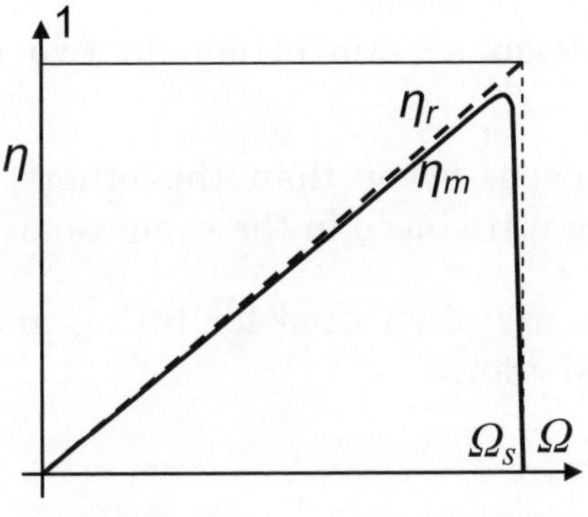

Figure 4.7. *Efficiency characteristics according to the speed η_r = rotor efficiency, η_m = overall motor efficiency*

4.3.4. *Two-speed motor – Dahlander's coupling*

An asynchronous motor presents as a rule a stabilized domain of speed varying by several percent. Outside this, the efficiency quickly decreases. It is however possible to realize a two-speed motor in a ratio of one to two with a single winding, by modifying the coupling of this last one. Two variants are possible:

– a triangle coupling with two ways in parallel being transformed into a star coupling in series (Figure 4.8);

– a star coupling with two ways in parallel being transformed into a triangle serial coupling (Figure 4.9).

The change of connection mode leads to an inversion of the current direction in half of the ways and so a change of polarity. Among both presented variants, the preference will go towards the one most frequently functioning in star.

This solution is known as Dahlander's coupling. The applications are generally in the ranges of several kW to several tens of kW.

Induction motors with three speeds also exist:

– two speeds in a ratio of one to two by Dahlander's winding type;

– a third speed, lower than the others, obtained by a specific winding introduced in the same slots.

Such motors are often used for lifting machines (cranes, hoists, overhead cranes).

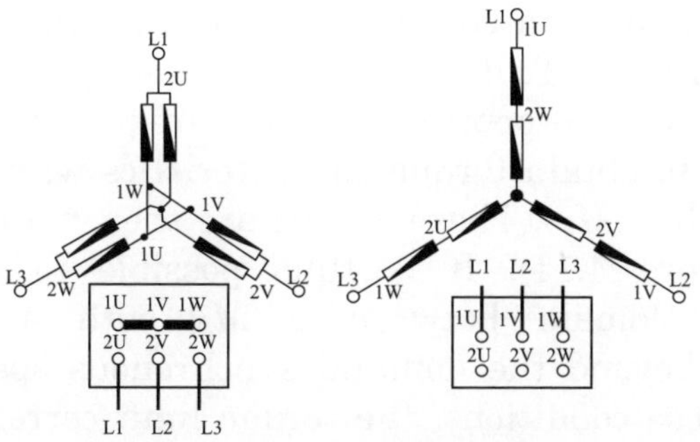

Figure 4.8. *Dahlander's coupling parallel triangle – star*

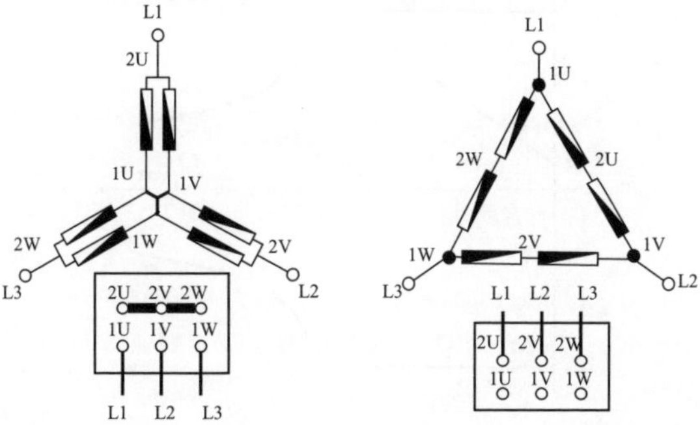

Figure 4.9. *Dahlander's coupling parallel star – triangle*

4.3.5. *Characteristics of current and torque with variable frequency*

Figure 4.7 brings to light the bad efficiency of the motor supplied with constant frequency at different speeds to those around the synchronous speed. In particular, for a speed of half the synchronous speed, the efficiency is lower than 50%. This is acceptable for a rarely occurring transient start-up phase, but is no longer acceptable for very frequent start-ups or for variable speeds (trains, trams, machine tools, etc.).

The main solution consists of supplying the motor with *variable frequency*. To do this, the power supply voltage must be approximately proportional to the frequency (see Figure 4.10), so as to obtain torque characteristics with a constant maximal value (M_K). These curves are then parallel to each others (Figure 4.11). It is thus possible to generate a functioning domain between ± M_N with a hyperbolic diminution beyond the nominal synchronous speed (Figure 4.10). In these conditions, the torque limit corresponds to a constant current equal to the rating value.

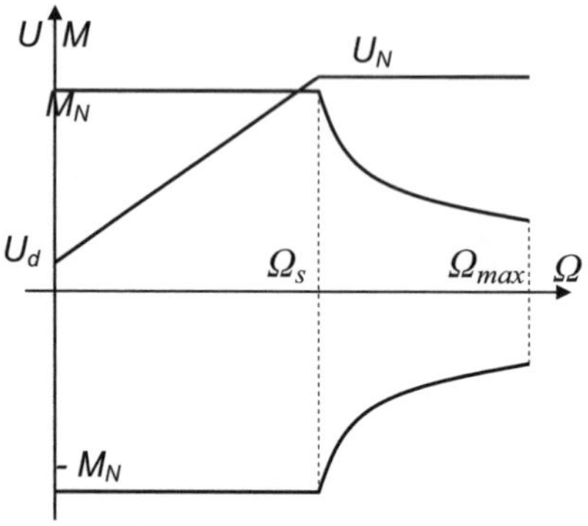

Figure 4.10. *Characteristics of voltage and torque with variable frequency according to the speed*

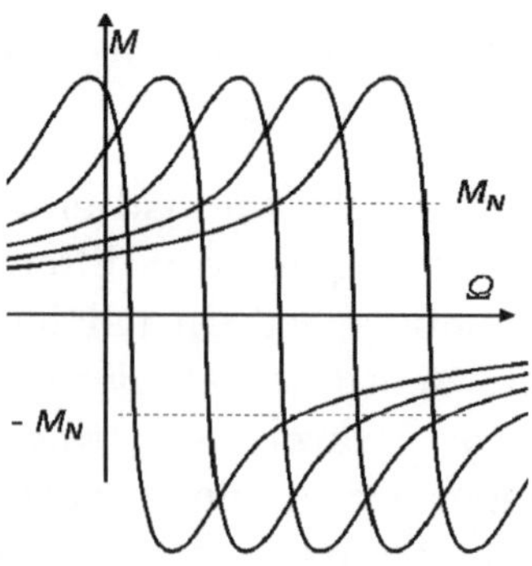

Figure 4.11. *Torque characteristics at different frequencies according to the speed*

4.3.6. *Wound induction motor*

The characteristic of asynchronous torque with constant frequency, presenting a value below the rating torque at standstill (power beyond several kW) and simultaneously a high current value, is not very favorable for heavy start-up, with high driven inertia. A solution consists of using a *wound rotor* with rings, with the possibility of inserting a variable resistance in the starting up (Figure 4.12). The stator is supplied by a three-phase source, the rotor has current running through it resulting from the induction phenomenon and presenting a variable resistance. The variable resistance is generally outside the motor and connected to the rotor winding through rings and through brushes (Figures 4.12 and 4.13).

Figure 4.12. *Wound asynchronous rotor with three pick up rings*

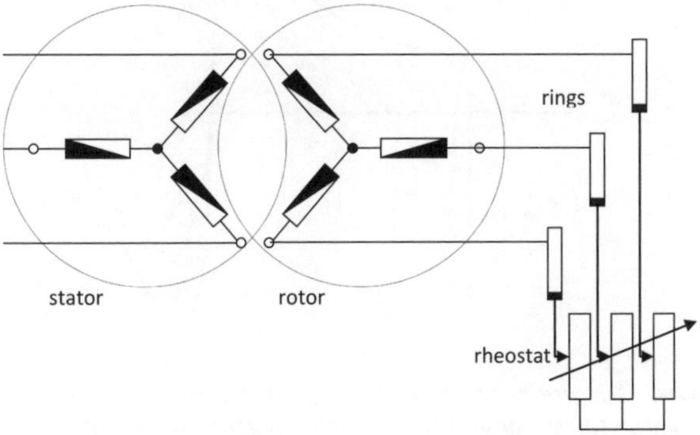

Figure 4.13. *Electric scheme of an induction motor with wound rotor, with variable resistance inserted by means of rings and brushes*

The torque characteristic presents a constant maximum amplitude, the slip of which is bound to the resistance by the expression:

$$s_K = \frac{R_r + R_d}{X_{cc}} \qquad [4.7]$$

with:

R_d the external starting resistance.

Figure 4.14 illustrates the torque evolution for diverse values of resistance.

Figure 4.15 describes the function starting current for diverse values of the resistance.

These two figures show that it is possible to increase the starting torque while reducing the current. Furthermore, a consequence of this solution, an important part of the rotor's start-up losses appear outside the motor.

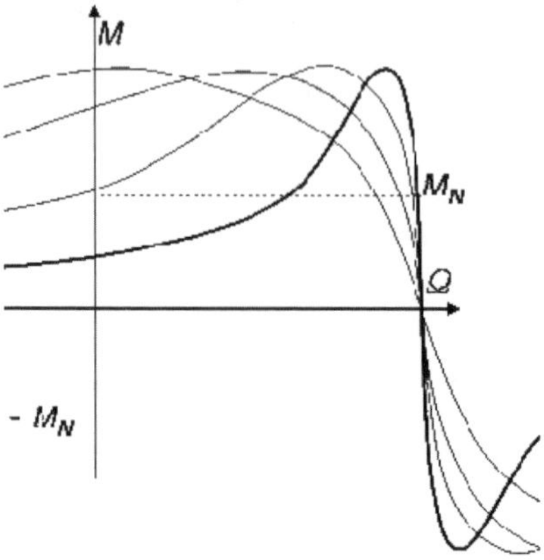

Figure 4.14. *Torque characteristics of an induction motor with wound rotor for diverse values of the of start-up resistance*

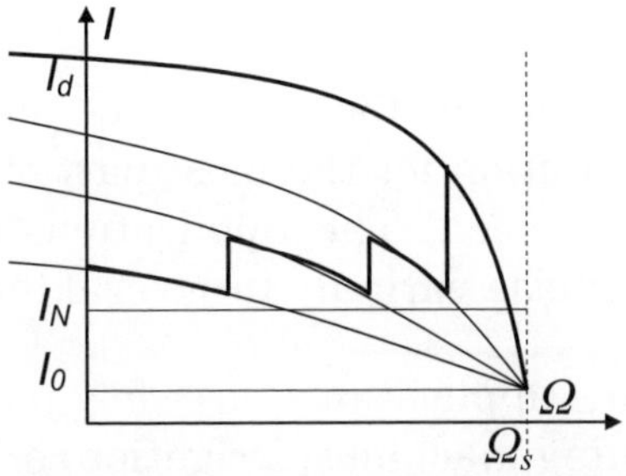

Figure 4.15. *Current characteristics of an induction motor with wound rotor for diverse values of the of start-up resistance*

4.3.7. *Single-phase induction motor*

For low powers (lower than 2 kW), we resort frequently to variants of *single-phase induction motors*.

The most frequent solution is in fact a two-phase motor for which one of the phases is supplied through a condenser (Figure 4.16a). This creates a phase shift about $\pi/2$ between the currents of both phases, such as described by the complex diagram of Figure 4.16b. With a sensible choice of the respective numbers of turns of both phases, it is possible to create a two-phase rotating field. However, this is true only for one slip, generally corresponding to the rated load.

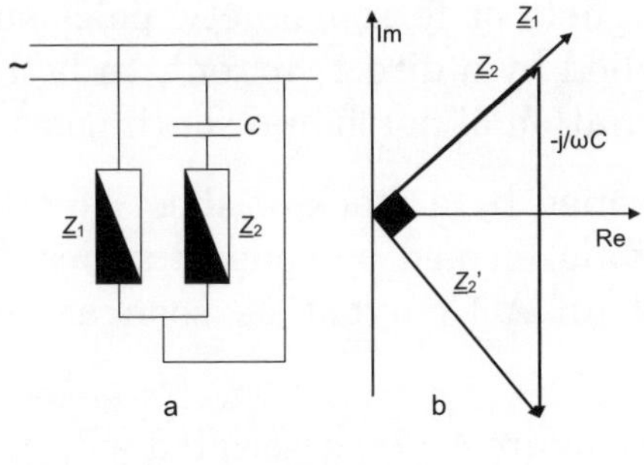

Figure 4.16. *a) Two-phase induction motor supplied in single-phase, with condenser; b) Complex graph with corresponding impedances*

4.3.8. *Application fields*

The asynchronous motor is, by its simplicity and its robustness, the solution for the most part of the applications of quasi-constant speed. The applications go from several Watts for single-phase variants to several tens of MW.

In low power, applications are for pumps (hot water circulation, washing machines, irrigation) or for ventilators.

In average power, applications are for pumps and ventilators, but also lifting machines, machine tools, compressors, etc.

In high power, the main applications are railway vehicles (streetcars, trolleybuses, locomotives) supplied with variable frequency, but also large pumps and big installations for ventilation (tunnels for example).

4.4. DC motors

4.4.1. *Structure*

The *DC motor* [JUF 95a] is characterized by the following elements:

– a stator formed by a ferromagnetic yoke, support of permanent magnets or ferromagnetic poles surrounded with windings supplied by a direct current. In both cases we thus create an alternation of north and south poles (Figure 4.17a);

– a rotor formed by a stack of sheet steels provided with slots in a winding formed by coils in series. The winding is the support of an AC current as soon as the rotor moves (Figure 4.17b);

– a *collector* (Figure 4.17c) associated with the rotor, formed by copper blades connected regularly with the rotor winding and on which *carbon brushes* rub. There are as many brushes (or brush rows) as there are poles in the stator.

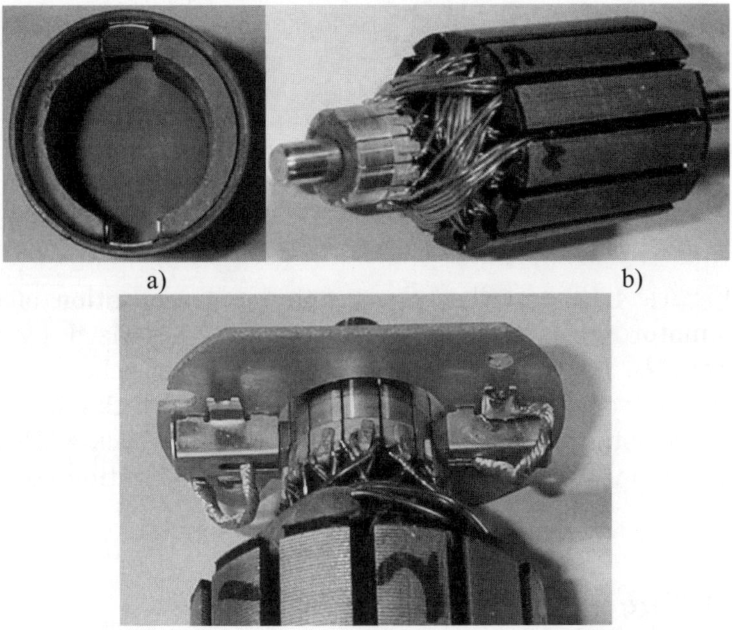

a) b)

c)

Figure 4.17. *a) Bipolar DC motor with permanent magnets – stator; b) DC motor – rotor; c) collector and coal brushes*

The role of the collector, associated with the rotor winding, is that of a rectifier (generator mode) or of a DC-AC converter (motor mode), by electromechanical switching.

4.4.2. *Equivalent electric scheme*

Figure 4.18 describes the graphic representation of the DC motor with permanent magnets as well as its equivalent electric scheme. This last one is mainly characterized by the internal resistance of the rotor R_r and by the movement back-EMF U_i, proportional to the rotating speed.

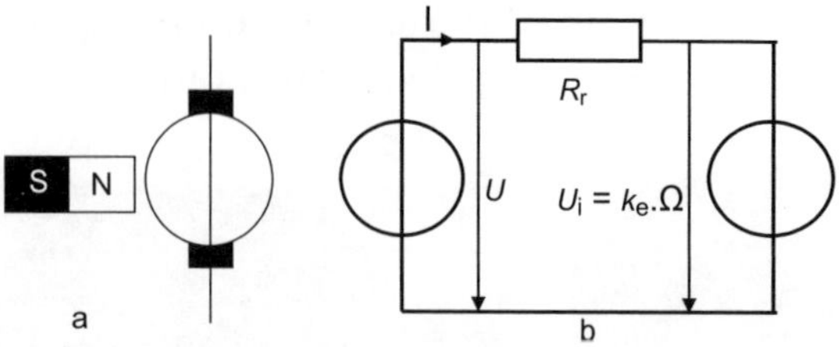

Figure 4.18. *a) DC motor with permanent magnets – graphic representation; b) equivalent electric scheme*

Figure 4.19 describes the graphic representation of the DC motor with separate excitation, characterized by the current I_e, as well as its equivalent electric scheme. This electrical scheme is mainly characterized by the internal rotor resistance R_r and by the movement back-EMF U_i, proportional to the rotating speed and the excitation current I_e.

4.4.3. *Torque and current characteristics*

In the case of the motor with permanent magnets, by referring on the equivalent scheme of Figure 4.18, the rotor current is:

$$I = \frac{U - k_e \Omega}{R_r} \qquad [4.8]$$

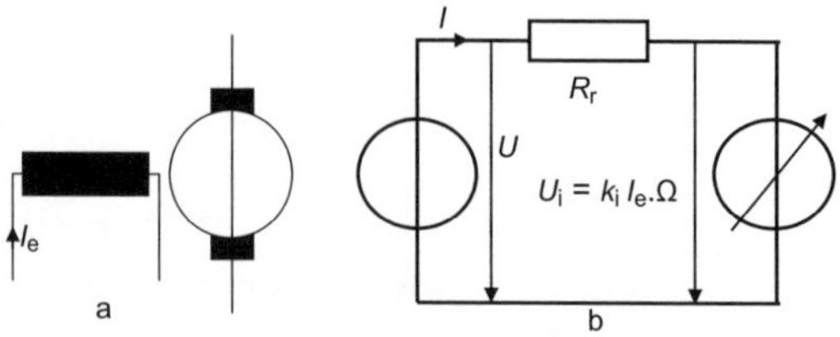

Figure 4.19. *a) DC motor with separate excitation – graphic representation b) equivalent electric scheme*

The corresponding torque is:

$$M = k_e I$$

Figure 4.20 shows the current and torque evolutions according to the speed, with a constant rotor voltage.

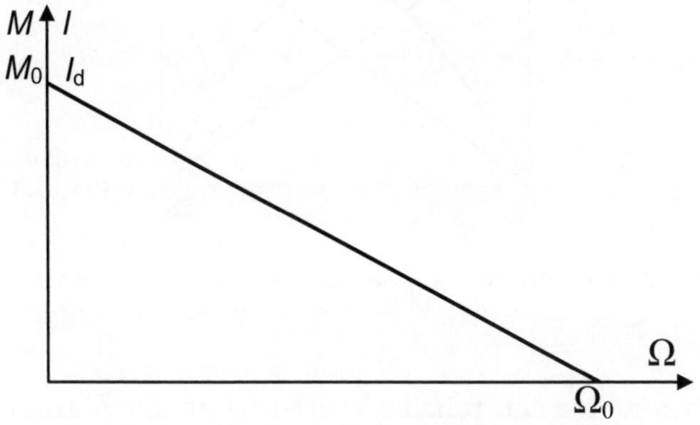

Figure 4.20. *Current and torque characteristic of the motor with permanent magnets with constant power supply voltage*

The corresponding mechanical power has the following expression:

$$P_{mec} = U_i I = k_e \Omega \frac{U - k_e \Omega}{R} \qquad [4.9]$$

It is thus a parabola according to the speed, presenting a maximum for $\Omega_0/2$ (Figure 4.21).

The efficiency corresponding to copper losses only is given by the following expression:

$$\eta_J = \frac{P_{mec}}{P_{el}} = \frac{U_i I}{UI} = \frac{k_e \Omega}{U} \qquad [4.10]$$

This efficiency has a linear variation with the speed and has a value of 1 at the no load speed Ω_0 (Figure 4.19).

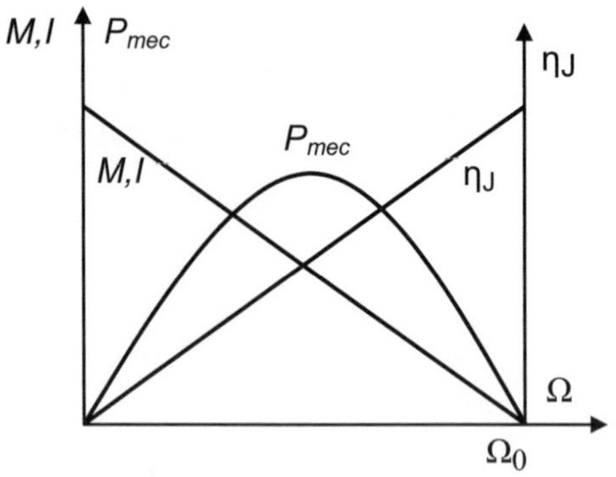

Figure 4.21. *Characteristics of torque, copper efficiency and mechanical power with constant voltage*

Such a motor can reach a start-up current I_d from 3 to 20 times the rated current, according to the value of the rotor resistance [4.8]. Consequently, for powers greater than several dozens of Watts, the current must be limited by an electronic device (controlled rectifier, chopper, etc.). It is thus possible to obtain a torque and current characteristic such as the one presented in Figure 4.22.

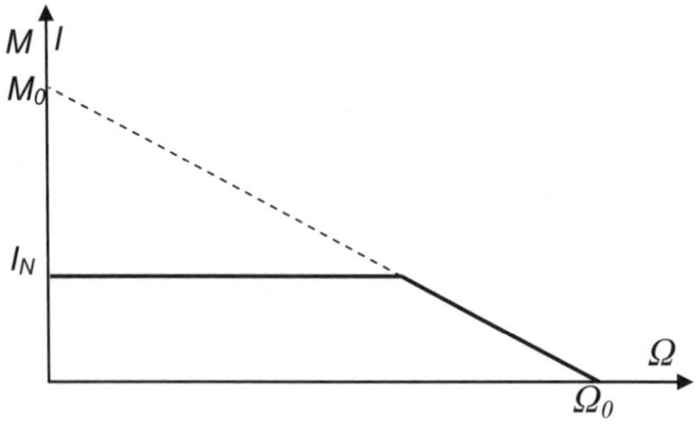

Figure 4.22. *Current and torque characteristic of a motor with permanent magnets, with current control by an electronic device*

During start-up or during braking with a constant current equal to the rated one, the cycle described in Figure 4.23 is followed, with a linearly increasing or decreasing voltage.

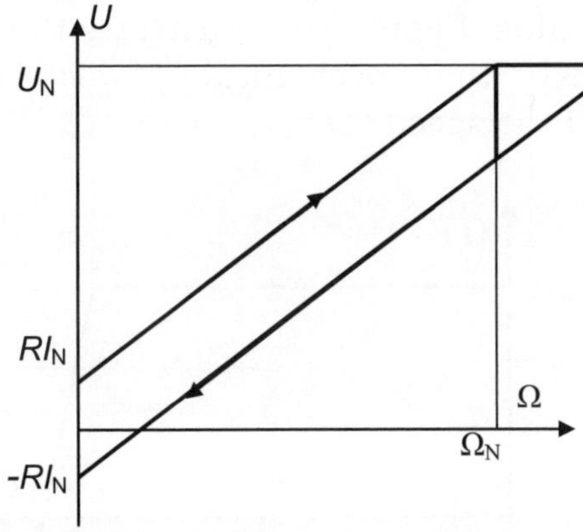

Figure 4.23. *Voltage characteristic with constant current during start-up or braking*

With this kind of current control, the functioning domain of the motor with permanent magnets becomes that of the hatched zone in Figure 4.24.

For a motor with excitation winding, generally with a power of several kW, an electronic control device for the current is indispensable.

The current and torque expressions are then:

$$I = \frac{U - k_e' i_e \Omega}{R_r} \qquad [4.11]$$

$$\Omega = \frac{U - R_r I}{k_e i_e'} \qquad [4.12]$$

$$M = k_e' i_e I \qquad [4.13]$$

For speeds lower than or equal to Ω_N, the excitation current is maintained at a maximal value (i_{emax}) so as to obtain a minimum of rotor current I. Beyond this, the excitation current is reduced so as to maintain the current I at its nominal value. Figure 4.25 illustrates this behavior. In these conditions, the mechanical power is practically constant beyond the speed Ω_N.

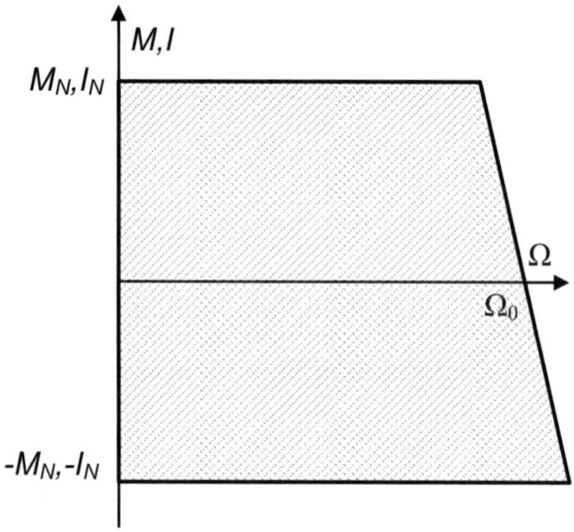

Figure 4.24. *Functioning domain in motor and generator with variable voltage*

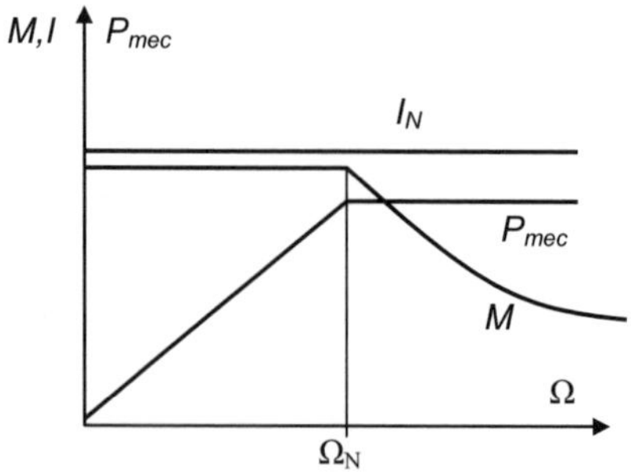

Figure 4.25. *Characteristics of current, torque and mechanical power of a n excitation DC motor, with current control by an electronic device*

4.4.4. *Collector motor*

The *collector motor* or *universal motor* is a motor with serial excitation supplied with AC current (Figure 4.26).

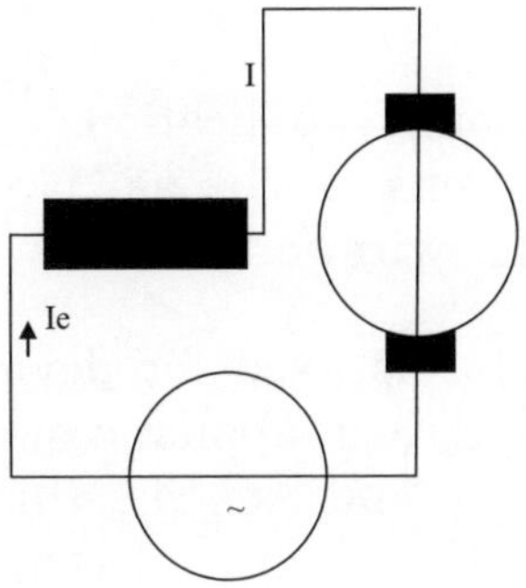

Figure 4.26. *Collector motor with serial excitation, supplied in AC current*

The torque can be written as:

$$M = k_e^{'} i^2$$

$$\overline{M} = k_e^{'} \frac{\hat{I}^2}{2} = k_e^{'} I^2 \qquad [4.14]$$

The speed can be adjusted by means of the supply voltage or by the introduction of a resistance in parallel with the stator winding.

4.4.5. *Applications*

The DC motor has applications which go from several fractions of Watts to several dozens of kW, more rarely up to some MW, generally for variable speeds.

For low powers, the applications are generally toys, or in the medical, automotive and micro-technology domains.

For average powers applications are, amongst others numerous electric tools, household electrical appliances, ventilators with variable flow, automotive applications (a car of average range has more than 50 motors or electric actuators), etc.

For major powers, applications are generally for production installations such as paper manufacturing, printing, rolling mills, wireworks, etc.

The collector motor is used for devices such as electric tools, household electrical appliances, hairdryers, etc. In Switzerland, Germany and Austria, this type of motor has been used for a long time for the electric railroad drive with a frequency of 16.66 Hz and powers beyond MW.

Increasingly, the DC-motor is tending to be supplanted by the self-commutated synchronous motor for high-performance applications, high reliability and long lasting life as well as for certain big series.

4.5. Synchronous motors

4.5.1. *Structure*

The *synchronous motor* [JUF 95a] is characterized by a stator supplied by AC, generally three-phase. The rotor can have three different structures:

– a rotor with permanent magnets (PM): Figure 4.27a illustrates three possible variants, while Figure 4.27b describes the structure with a poly-phase stator;

– a rotor equipped with an excitation winding and with salient poles (Figure 4.28), identical to that of a synchronous alternator;

– a ferromagnetic rotor with a tooth-shaped pole structure, without winding. It is called *variable reluctance motor* or *reluctant motor* (Figure 4.29).

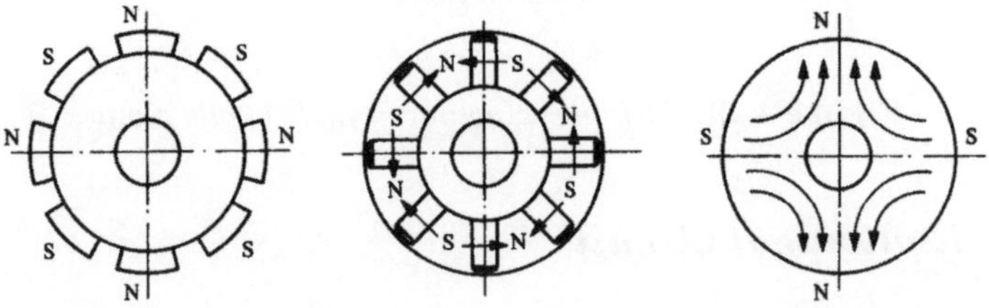

Figure 4.27. *a) Rotor structures of PM synchronous motors*

Figure 4.27. *b) PM synchronous motor structure – stator and rotor*

Figure 4.28. *Rotor of a synchronous motor-alternator with 6 salient poles of 45 MW for pump-turbine installation. Also see Figure 7.10*

Figure 4.29. *Variable reluctance synchronous motor*

4.5.2. *Equivalent circuit*

In sinusoidal supply mode and back-EMF voltage, the synchronous motor is characterized by the following phase voltage equation:

$$u_s = R_s i_s + L_s \frac{di_s}{dt} + k_e \Omega \sin \omega t \qquad [4.15]$$

In complex writing, the same phase equation is written:

$$\underline{U}_s = (R_s + j\omega L_s)\underline{I}_s + k_e\Omega = \underline{Z}_s \underline{I}_s + k_e\Omega \qquad [4.16]$$

$$U_e = k_e\Omega = \text{movement back-EMF} \qquad [4.17]$$

The *equivalent circuit* in Figure 4.30 represents this voltage equation for a phase.

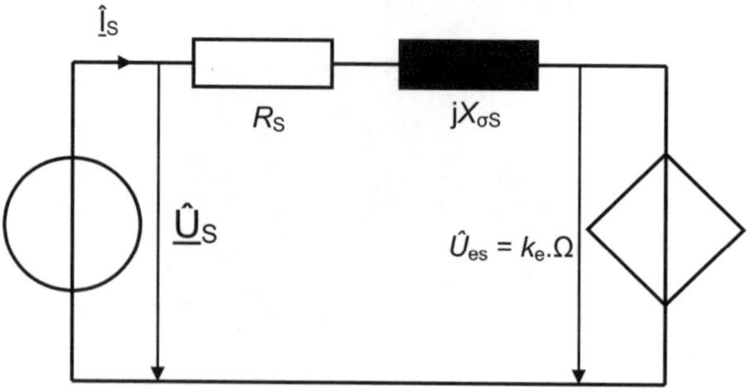

Figure 4.30. *Equivalent circuit for a phase of a PM synchronous motor*

Figure 4.31 describes the *vector diagram* corresponding to the equivalent circuit of Figure 4.30 as well as the associated angles.

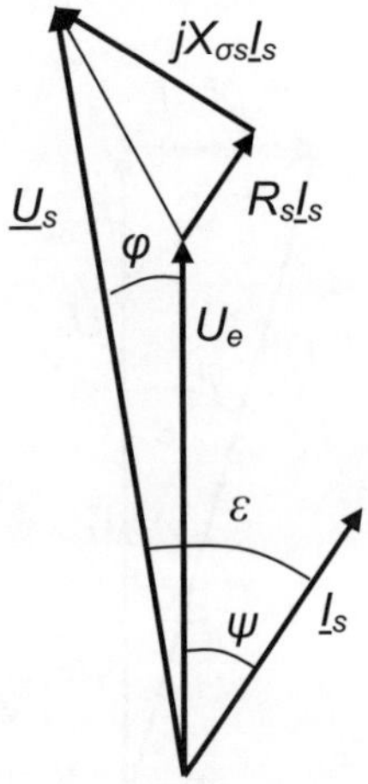

Figure 4.31. *Voltage and current vector diagram for a phase of a PM synchronous motor*

4.5.3. *Torque characteristics with imposed current*

For a three-phase motor with permanent magnets, the torque expression is:

$$M = \frac{3}{\Omega} U_e I_s \cos\psi = 3k_e I_s \cos\psi \quad [4.18]$$

It is possible to create a maximum torque for a given current by cancelling the ψ angle corresponding to the phase shift between the current and the back-EMF voltage. Figure 4.32 illustrates the voltage and current vector diagram for this case.

$$\hat{M} = 3k_e I_s = k_M I_s \quad [4.19]$$

To do this, it is necessary to control the motor supply by means of a sensor placed on the motor shaft, and delivering a voltage in phase with the back-EMF voltage (see section 10.1).

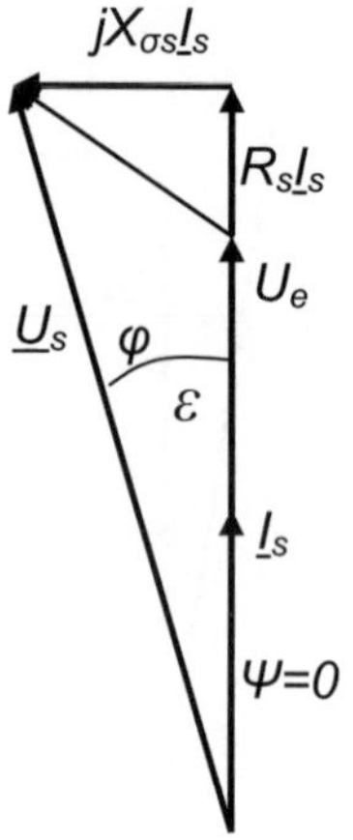

Figure 4.32. *Voltage and current vector diagram for the current in phase with the back-EMF*

4.5.4. *Torque characteristics with imposed voltage*

By referring to equation [4.16], the current expression, according to the supply voltage and to the back-EMF, is worth:

$$\underline{I}_s = \frac{\underline{U}_s - \underline{U}_e}{\underline{Z}_s} \qquad [4.20]$$

By replacing [4.18] in the torque expression and by developing, the torque for three phases is:

$$M = 3\frac{k_e}{Z_s}\left[U_s \cos(\varphi_s - \varepsilon) - U_e \cos\varphi_s\right] = M_{mut} - M_f \qquad [4.21]$$

The term function of the angle ε corresponds to the mutual interaction between magnets and current: it is positive or negative. The term M_f is always negative (braking effect). This is due to the magnet only.

Thus the torque varies with a sine shape according to ε (Figure 4.33). It presents a maximum and a minimum. For the maximum, with $\varepsilon = \varphi_s$:

$$\hat{M} = 3\frac{k_e}{Z_s}\left[U_s - U_e \cos\varphi_s\right] \qquad [4.22]$$

For the minimum, with $\varepsilon = \varphi_s + \pi$:

$$\breve{M} = 3\frac{k_e}{Z_s}\left[U_s + U_e \cos\varphi_s\right] \qquad [4.23]$$

The angle ε adapts itself according to the load torque M_{res}. Figure 4.33 shows a crossing point of stable balance (angle ε_1) for a given load torque.

$$M = 3\frac{k_e}{Z_s}\left[U_s \cos(\varphi_s - \varepsilon) - U_e \cos\varphi_s\right] = M(\varepsilon) = M_{res} \qquad [4.24]$$

4.5.5. *Self-commutated mode*

The synchronous motor with permanent magnets is generally supplied in *self-commutated mode* by an electronic supply, piloted by a rotor *position sensor* (see section 10.1). This means imposing the angle ε on a value ε_1. The adjustment between motor torque and load torque is then made by means of the speed Ω. Equation [4.24] becomes:

$$M = 3\frac{k_e}{Z_s}\left[U_s \cos(\varphi_s - \varepsilon_i) - U_e \cos\varphi_s\right] = M(\Omega) = M_{res}$$

[4.25]

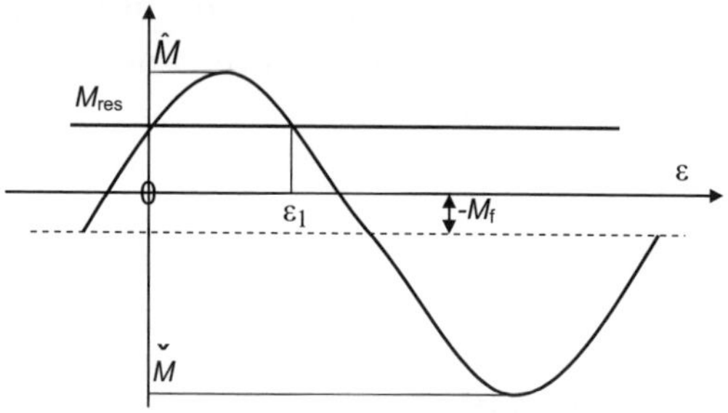

Figure 4.33. *Torque evolution according to the angle ε*

The quantities Z_s, U_e and φ_s are all functions of speed and adapt to ensure a balance between the two torques. If this is not the case, motor speed changes up or down until this balance is achieved.

This type of motor is usually called a *brushless DC motor (BLDC motor)*. Compared to the DC motor, in addition to the permutation of electric and magnetic components between stator and rotor, winding switching is ensured by an electronic converter instead of a collector.

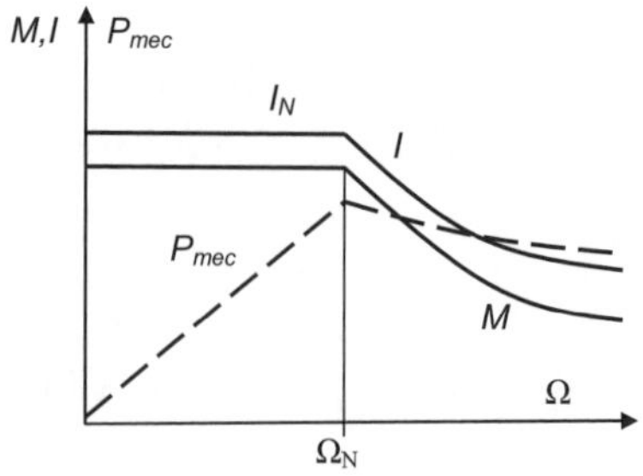

Figure 4.34. *Current characteristics torque and power of a brushless DC motor*

For most applications, the BLDC motor works with an imposed current until the nominal speed set by the rated voltage, then with imposed voltage and controlled angle ε. Figure 4.34 shows the evolution of current, torque and mechanical power under these conditions.

Figure 4.35 describes this kind of motor operating area. Brake (generator) and motor limits are non-symmetrical, according to the term M_f given by Figure 4.33.

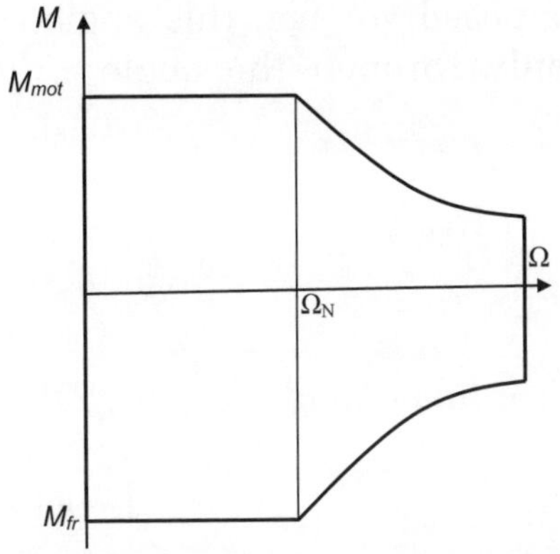

Figure 4.35. *Operation area of a BLDC motor*

4.5.6. *Angle ε adaptation for a constant voltage supply*

Possibilities for control of a BLDC motor will be discussed with an example described in the appendix for this chapter (section 4.9).

For a constant voltage supply, Figure 4.36 describes the torque characteristics of this motor for different speeds, depending on angle ε. Amplitude decreases with speed. This is mainly linked to a decreasing current due to the impedance Z_s increasing. Furthermore, the sine torque shifts to increasing negative values of ε, with the speed increasing. At standstill, the angle ε leading to the maximum torque is zero, depending on the diagram in Figure 4.32. At a speed of 500 rpm, this optimum appears for an angle of approximately -75°. If the aim is to impose a maximum torque for any speed, this angle should evolve by creating a phenomenon of commutation-advance similar to spark-ignition advance in explosion motors. This corresponds to the ε_m curve of Figures 4.36 and 4.37.

The best use of a BLDC-motor is to obtain the best torque/current ratio (k_M), which is a zero ψ angle (see section 4.5.3). With imposed voltage, this angle cannot be imposed directly but only through the angle ε. The corresponding value is:

$$\varepsilon_I = \arccos\left(\frac{U_e \omega L_s}{Z_s U}\right) - arctg\left(\frac{R_s}{\omega L_s}\right) \qquad [4.26]$$

This angle is a function of speed (ω, Z_s) and voltage U.

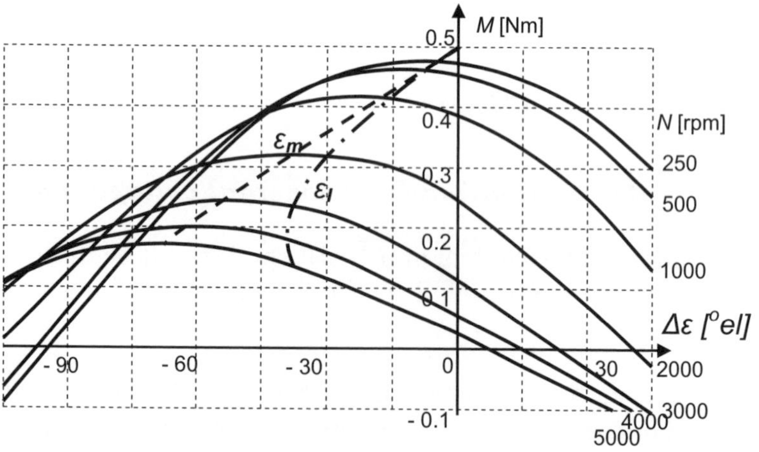

Figure 4.36. *Torque characteristics depending on the angle ε for different speeds at constant voltage for the motor in the appendix (section 4.9)*

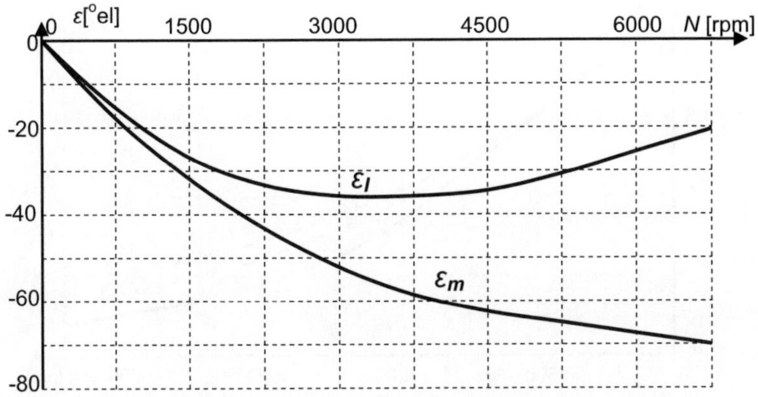

Figure 4.37. *ε angle evolution depending on the speed for a constant voltage; ε_m for a maximum torque; ε_I for the best ratio power / torque (motor in the appendix, section 4.9)*

Figures 4.38 and 4.39 respectively describe the evolution of the torque and current for three control modes of angle ε:

– ε constant equal to 0 (M_0, I_0);

– ε_m for a maximum torque (M_m, I_m);

– ε_I for a ratio torque/current optimum (M_I, I_I).

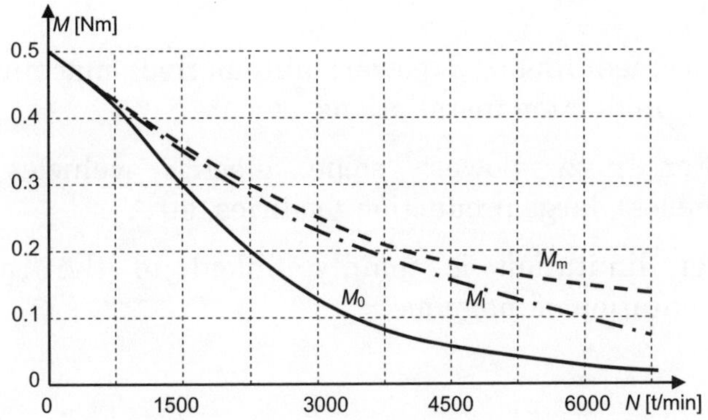

Figure 4.38. *Torque evolution depending on the speed with constant voltage for 3 control modes: M_0 for an angle ε = 0; M_m for the maximum torque; M_I for the best ratio torque / current*

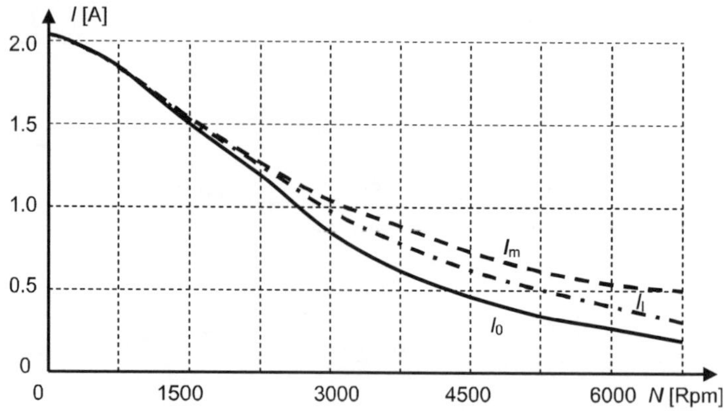

Figure 4.39. *Current evolution depending on the speed with constant voltage for 3 control mode: I_0 for an angle $\varepsilon = 0$; I_m for the maximum torque; I_l for the best ratio torque / current*

4.5.7. *Applications*

The self-commutated synchronous motor or BLDC motor has applications from several μW to several dozens of kW, and more rarely beyond, generally for variable speeds:

– For low power: applications watch and clock drives, peripheral devices for computers, medical and automotive domains.

– For medium-sized power: automotive, machine tools, robotics, production installations, etc.

– For high power: some electric vehicles (cars, trolleybuses), large production facilities, etc.

Power limitation is mainly linked to the cost and implementation of magnets.

4.6. Variable reluctance motors

4.6.1. *Structure and characteristics*

Figure 4.40 describes a typical example of a *variable reluctance motor* [JUF 95a] with unipolar winding. The torque created by one phase of such a motor has the following expression:

$$M_{ph} = \frac{1}{2} \frac{\partial L_{ph}}{\partial \alpha} i_{ph}^2 \quad [4.27]$$

The torque sign is therefore not influenced by the sign of the current. The torque is thus proportional to the current square without saturation. In reality it tends towards a linear characteristic for high current values (Figure 4.41).

The inductance presents a maximum when the rotor poles are opposite to the corresponding powered phase stator teeth. Similarly, the inductance presents a minimum when the inter-polar rotor area is opposite to the corresponding powered phase stator teeth.

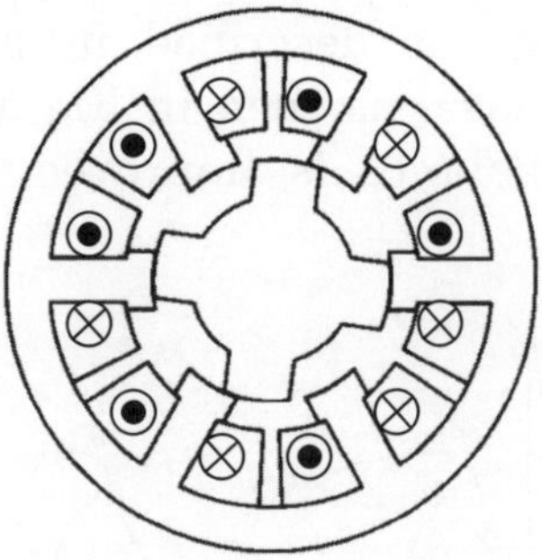

Figure 4.40. *Variable reluctance motor, with 3 phases, 6 teeth and 4 rotor poles with winding*

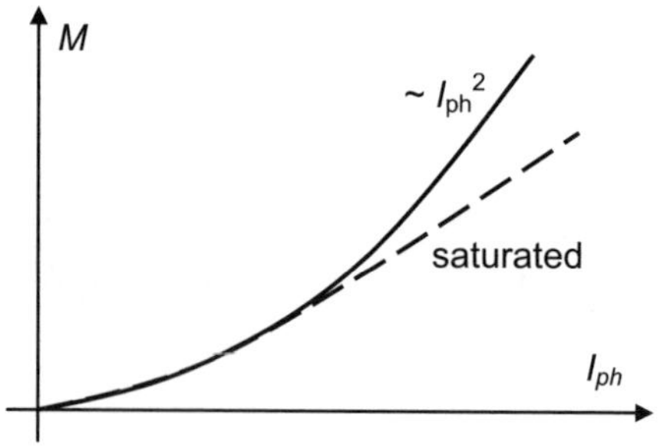

Figure 4.41. *Variable reluctance motor, peak torque evolution with current level*

4.6.2. *Driver and applications*

Figure 4.42 describes a torque shape depending on the position for a three-phase motor phase, for which phases are successively supplied at constant current. By switching according to the position, it is possible to create a constantly positive (or negative) torque. It is also possible to supply two phases simultaneously, allowing us to increase the torque.

To be effective, this type of motor shall be supplied in unipolar self-commutated mode (one-way current). It then uses an electronic bridge as described in Figure 9.3 or three bridges (one per phase) as seen in Figure 9.6. In the latter case, the energy balance is improved and the maximum speed is increased.

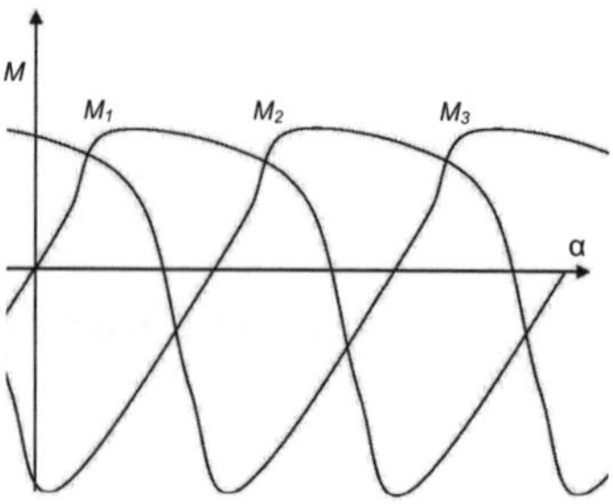

Figure 4.42. *Torque characteristic of a switched reluctance motor*

Due to its small air gap, the important torque harmonic level and a switch mode creating discontinuities, the reluctance motor is noisy. Its applications are themselves therefore noisy themselves as tools (drills, grinders, electric hammers, etc.), compressors, grinders and generally for drives at high speeds.

4.7. Linear motors

4.7.1. *Induction motors*

Figure 4.43 shows a *linear induction motor* structure including an induction armature built of a continuous conductive material (aluminum) and a ferromagnetic yoke to close the magnetic flux.

Figure 4.44 shows a linear induction motor with double inductor located on both sides of a conductive aluminum armature.

The main disadvantage of the linear induction motor is an efficiency reduction as a result of a more significant air gap than for a rotating motor and the electric and magnetic armatures which are superimposed instead of integrated.

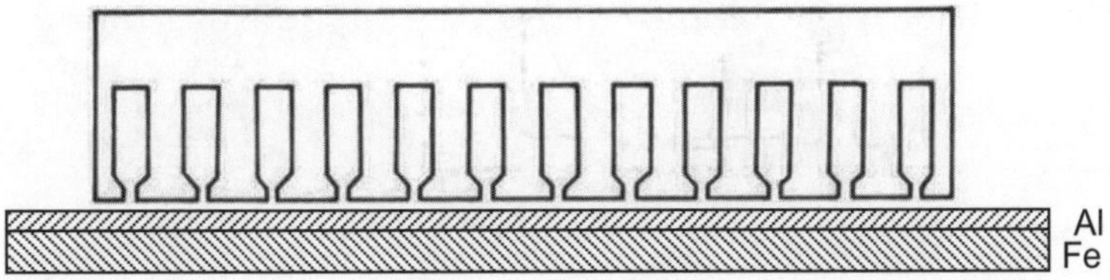

Figure 4.43. *Structure of an asynchronous linear motor with a single armature in Al and Fe*

Figure 4.44. *Linear induction motor with double structure around a conducting armature*

Figure 4.45 shows the advantage of using linear induction motors for the propulsion of metro trains to reduce the tunnel diameter by lowering the level of the vehicle floor. The result is a significant reduction of investment costs. This technology is specially developed in Japan.

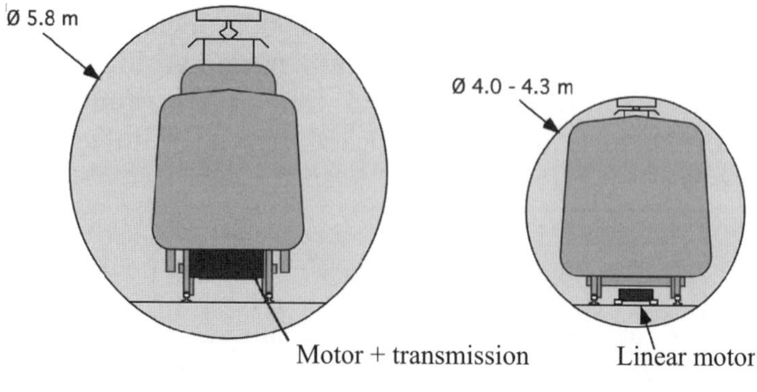

Figure 4.45. *Classic and linear motor variants for metro vehicles*

4.7.2. *Synchronous motors*

Variants of *linear synchronous motors* or *BLDC motors* are also possible. Figure 4.46 shows the structure of such a motor. Figure 4.47 presents the assembled system. The main applications are in the field of machine tools which can achieve better accuracy and rigidity than a rotary motor linked to a ball screw. Figure 4.1b shows an example.

4.7.3. *Moving coil motors*

Applications requiring a high motion dynamic with a reduced amplitude frequently resort to motors or linear actuators with a *moving coil*. Figure 4.48 shows a typical structure. An E-shaped structure consists of a ferromagnetic material either only slightly or unsaturated. It supports two permanent magnets at high energy density. The mobile coil is thus placed in a constant induction air gap. The result is a force proportional to the current and a direction linked to its sign.

By referencing [JUF 95a, Chapter 3], the following expressions for the gap induction and for the force, assuming an ideal ferromagnetic circuit are:

$$B_\delta = \frac{B_0}{1 + \mu_{dr}\delta / e} \qquad [4.28]$$

with:

B_0 = PM remnent flux density;

B_δ = air gap flux density;

μ_{dr} = PM internal relative permeability;

δ,e,a,c; see Figure 4.48.

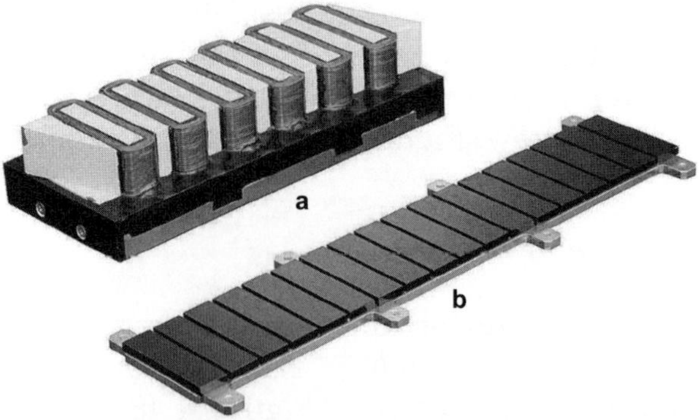

Figure 4.46. *Components of a linear synchronous motor with permanent motors: a) polyphased structure; b) track with permanent magnets*

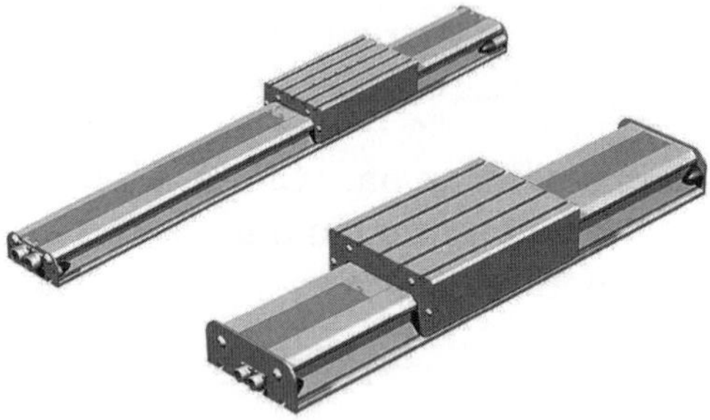

Figure 4.47. *Synchronous linear motors with PM*

The force on the coil is:

$$F = 2aNiB_{\delta} \quad [4.29]$$

with:

N = coil turn number

i = coil current

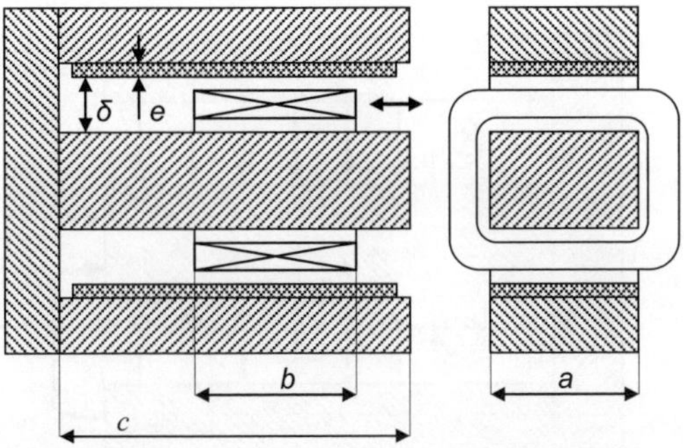

Figure 4.48. *Structure of a linear actuator with moving coil*

In fact, for the structure in Figure 4.48, a dissymmetrical force appears between positive and negative currents. This is attributable to a reluctance effect related to the axial dissymmetrical magnetic circuit. Figure 4.49 defines the axial position of the coil. By replacing it with an equivalent coil concentrated at the x coordinate, the field distribution created by the single coil current is represented in Figure 4.50.

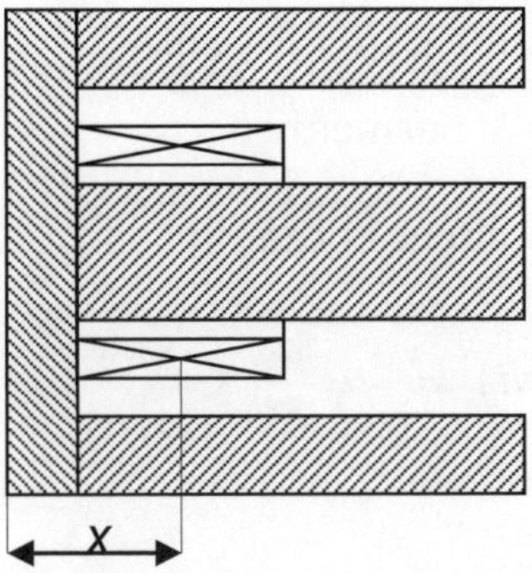

Figure 4.49. *Coil position*

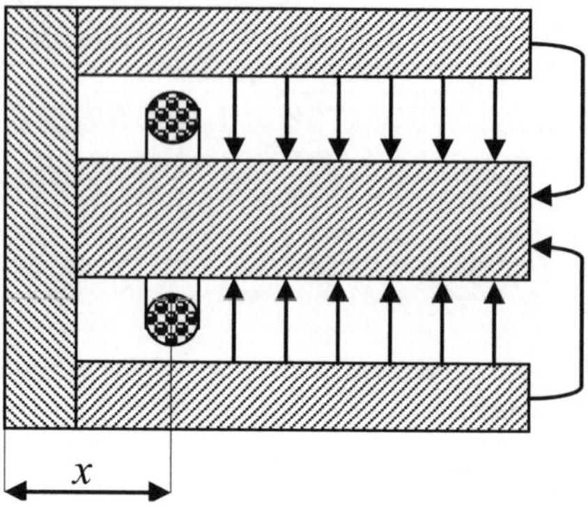

Figure 4.50. *Air flux density created by the coil*

Always assuming an ideal ferromagnetic circuit, the single coil permeance can be written as:

$$\Lambda_b = \mu_0 \frac{2a(c-x)}{\delta'} + \Lambda_{extr} \qquad [4.30]$$

with:

δ' = equivalent air gap = $\delta + e/\mu_{dr}$

Λ_{extr} = extremity permeance

The force of this effect is then:

$$F_x = \frac{1}{2}\frac{d\Lambda_b}{dx}(Ni)^2 = -\mu_0 \frac{a}{\delta'}(Ni)^2 \qquad [4.31]$$

$$F_{tot} = F + F_x$$

Figure 4.51 shows the resulting dissymmetrical forces. This default can be completely eliminated by a symmetrical magnetic circuit in accordance with Figure 4.52.

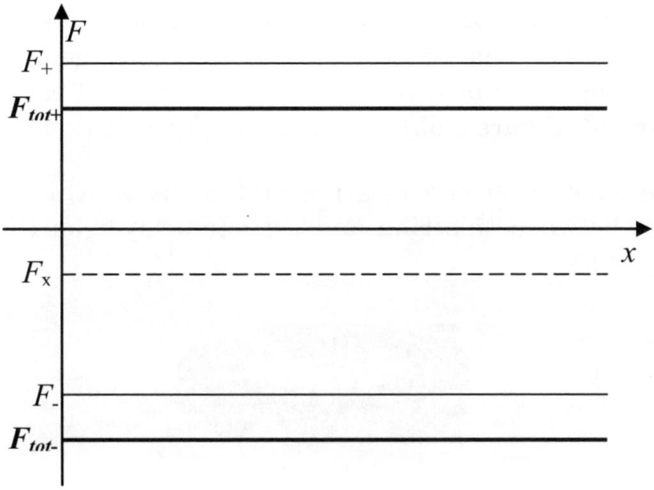

Figure 4.51. *Dissymmetrical forces created by the reluctance effect F_x*

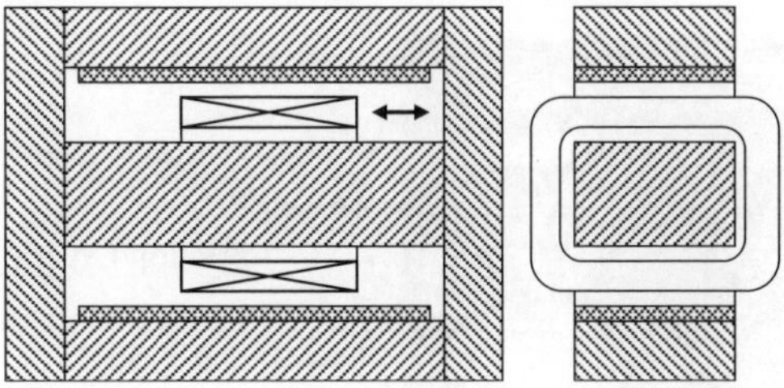

Figure 4.52. *Elimination of the reluctance effect of reluctance by a symmetrical ferromagnetic circuit*

4.8. Piezoelectric motors and actuators

4.8.1. *Piezoelectric motors*

Generally, piezoelectric systems are based on the property of non-conductive materials generating an electric field when they are subjected to mechanical stress. Conversely, if these materials are subject to a potential difference in a privileged axis, as a result a mechanical strain is created. It can be amplified by mechanical resonance.

In particular, a ring supporting peripheral sectors to which a polyphased voltage is applied generates a progressing wave in this ring. The movement of each sector is elliptical (Figure 4.53).

The generation of a rotating movement is achieved by a contact between this ring and a friction layer on the rotor (Figure 4.54).

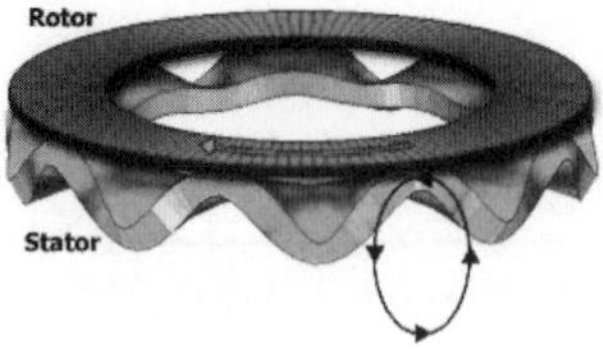

Figure 4.53. *Generation of a local elliptical movement on a piezoelectric ring*

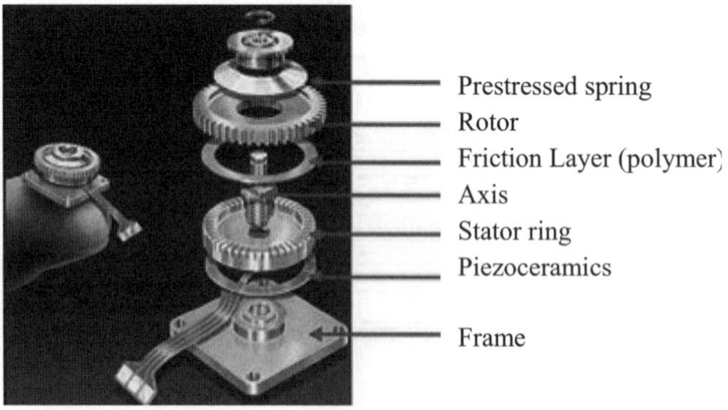

Figure 4.54. *Structure of a piezoelectric rotating motor*

The characteristic of such a motor is represented in Figure 4.55. It presents a standstill torque M_d, which allows us to lock a position without power. Adjustable settings are voltage (torque amplitude) and frequency (rotation speed).

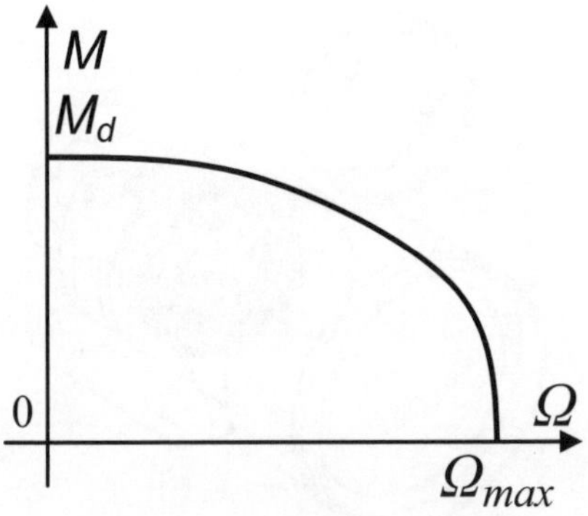

Figure 4.55. *Torque-speed characteristic of a rotating piezoelectric motor*

4.8.2. *Applications*

Small motors and actuators can be designed for robotics applications, which require relatively low power but high accuracy and limited weight (Figure 4.56).

Figure 4.56. *Piezo-motor for robotics*

The annular piezoelectric motor described in Figure 4.53 is applied to the zoom control of some cameras (Figure 4.57). The chosen position is automatically locked.

Figure 4.57. *Piezoelectric zoom motor for camera*

Automotive applications grow gradually, thanks to the position locking property and the silence of the movement. In the future, the piezoelectric motors will move windows, seats, centralized openings, valves, etc. A new application that should be coming into practice in the coming years is direct petrol injection by a linear actuator combining high pressure (200 bar) and high-speed (Figure 4.59).

4.8.3. *Piezoelectric activators*

A *piezoelectric actuator* is characterized by the creation of a linear motion with a large force and low movement. A mechanical amplification can be performed (Figure 4.58).

Applications include positioning systems, valve control, MEMs, etc.

Resonant systems can be used for the diffusion of liquids, gas injection for example (Figure 4.59).

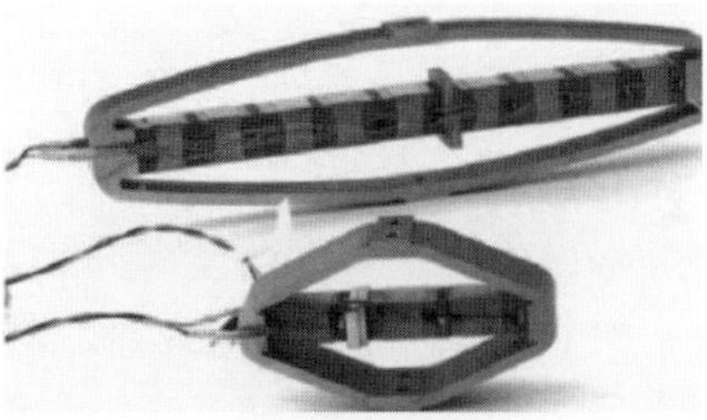

Figure 4.58. *Piezoelectric linear actuators with amplitude amplification*

Figure 4.59. *Piezoelectric petrol injection system*

4.9. Appendix – BLDC motor characterisitcs

This motor has the following characteristics:

Rated speed:	3,000 rpm
Rated power:	12 W
Pole number:	8
Rated voltage:	36 V (phase rms)
Phase resistance:	0.75 Ω
Phase inductance:	0.49 mH
Back EMF coefficient:	0.0162 Vs/rad (phase peak)

Chapter 5

Motors: Characterization

5.1. Characteristics

5.1.1. *Typology*

There are several ways to classify the different motor types. Focusing on the main ways, two characteristics mainly apply:

– according to the motor principle:

- DC motor,
- induction motor,
- synchronous motor,
- BLDC motor,
- reluctance motor;

– according to the excitation mode:

- DC excitation,
- excitation by induction,
- PM excitation,

- without (reluctance motor).

This final approach will be classified into two broad motor categories and set their dimensional performances.

5.1.2. *Aim of the scaling laws*

To characterize a motor associated with an electric drive, its main performances are the torque, the inertia, and the thermal constraints.

The *scaling laws* [JUF 95a], [JUF 96], rather than fixing the absolute characteristics of a motor, define the characteristic evolutions for a given motor type, depending on the relative sizes. To do this, a motor with a diameter d and a length ℓ is assumed to be known. The aim is to infer performances of a second motor defined by homothety from the first motor, with diameter d' and length ℓ' (Figure 5.1).

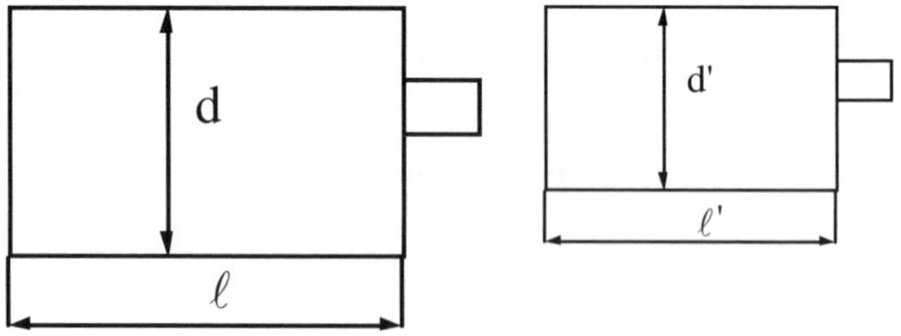

Figure 5.1. *Similar motors by homothety*

The second motor is characterized by the *scaling ratio* r* with:

$$r^* = \frac{d'}{d} = \frac{l'}{l} \qquad [5.1]$$

Any characteristic quantity *F'* of the second motor can be expressed in terms of the corresponding characteristic *F* of the first motor as:

$$F^* = \frac{F'}{F} = f(r^*) \qquad [5.2]$$

This approach will be applied to various motor types classified into different categories based on the excitation mode.

5.2. Scaling laws

5.2.1. *Copper losses*

Copper losses may be written as losses per volume unit:

$$P_J = \int_V \rho J^2 dV \qquad [5.3]$$

with, in relative values:

$\rho^* = 1$ same material (copper);

$V^* = r^{*3}$ relative volume. [5.4]

As a result:

$$P_J^* = J^{*2} r^{*3} \qquad [5.5]$$

5.2.2. *Resistances and inductances*

A winding resistance can be written as:

$$R = \rho \frac{l_{co} N}{S_{co} / N} \qquad [5.6]$$

with:

N = turn number in series;

l_{co} = turn length;

S_{co} = total copper section.

$$R^* = N^{*2} / r^* \qquad [5.7]$$

For an inductance:

$$L = N^2 \Lambda \qquad [5.8]$$

$$\Lambda = \int \mu \frac{dS}{l} = \text{permeance} \qquad [5.9]$$

$$\Lambda^* = r^* \qquad [5.10]$$

$$L^* = N^{*2} \, r^* \qquad [5.11]$$

It then becomes possible to determine the relative electrical time constant τ_{el}:

$$\tau_{el} = \frac{L}{R}$$

$$\tau_{el} = \frac{L^*}{R^*} = r^{*2} \qquad [5.12]$$

5.2.3. *Heating*

The motor heating at steady state can be defined with an equivalent convection [7.11] with:

$$\Delta\Theta = \frac{P_J}{\alpha_c S_c} \qquad [5.13]$$

with:

α_c = equivalent convection coefficient, with the same cooling mode:

$$\alpha_c{}^* = 1 \qquad [5.14]$$

S_c = global convection surface

$$\Delta\Theta^* = \frac{P_J{}^*}{r^*} \qquad [5.15]$$

Using relation [5.5]:

$$\Delta\Theta^* = J^{*2}\, r'' \qquad [5.16]$$

The objective is to use a motor on its power limit, or at its maximum heating. For similar motors, the insulation class is the same, hence it follows that:

$$\Delta\Theta^* = 1 \qquad [5.17]$$

$$J^* = \frac{1}{\sqrt{r^*}} \qquad [5.18]$$

5.2.4. *Induction and reluctance motors*

Both motor types are characterized by a torque proportional to the current square for a given speed. In particular, for a variable reluctance motor torque, the torque expression is:

$$M_{rel} = \frac{1}{2}\frac{d\Lambda}{d\alpha}\Theta^2 \qquad [5.19]$$

$$\Theta = Ni = JS_{cu}$$

$$\Theta^* = J^* r^{*2} \qquad [5.20]$$

$$\alpha^* = 1$$

Including [5.10] and [5.20]:

$$M_{rel}^* = J^{*2} r^{*5} \qquad [5.21]$$

By imposing constant heating ([4.18]), the relative torque is:

$$M_{rel}^* = r^{*4} \qquad [5.22]$$

It is then possible to determine the copper losses relating to the useful power using relations [5.5] and [5.21]. The mechanical power is:

$$\begin{aligned} P_{mec} &= M\Omega \\ P_{mec}^* &= J^{*2} r^{*5} \Omega^* \end{aligned} \qquad [5.23]$$

$$\frac{P_J^*}{P_{mec}^*} = \left(\frac{P_J}{P_{mec}}\right)^* = \frac{1}{r^{*2}\Omega^*} \qquad [5.24]$$

Regardless of the choice of the current density, the copper losses are increasing, so inversely proportional to the square of the homothety ratio. In other words the relative copper efficiency decreases rapidly with the size. It can be expressed as follows:

$$\eta_J' = \frac{1}{1+(P_J/P_{mec})^*(1/\eta_J - 1)} = \frac{1}{1+(1/\eta_J - 1)/(r^{*2}\Omega^*)} \qquad [5.25]$$

Efficiency does not depend on the current density, but only on the motor size. It is impossible to achieve a small reluctance motor with a good efficiency. It is the same for induction motors whose torque is also proportional to the current square.

5.2.5. *Permanent magnet motors*

Development that follows can be applied to any motor type whose excitation is created by one or more permanent magnets: DC-motor, synchronous motor, step motor, BLDC motor, etc.

The torque of a PM motor is proportional to the current with:

$$M_{ai} = \frac{d\Lambda_{ai}}{d\alpha}\Theta_{ai}\Theta_{i} \qquad [5.26]$$

Λ_{ai} is the mutual permeance between PM and winding; $\bullet_{ai}$ is the PM magnetic potential (or MMF) with [JUF 95, Chapter 3]:

$$\Theta_{ai} = H_0 l_{ai} \qquad [5.27]$$

H_0 is the coercive magnetic field of the PM recoil line [JUF 95]; for identical materials:

$$H_0{}^* = 1$$

$$\Theta_{ai}{}^* = r^* \qquad [5.28]$$

Using equations [5.10], [5.26] and [5.28], the relative torque is:

$$M_{ai} = J^* r^{*4} \qquad [5.29]$$

For constant heating, using [5.18], this expression becomes:

$$M_{ai} = r *^{3.5} \qquad [5.30]$$

The copper losses/mechanical power ratio, using [5.5] and [5.30] becomes:

$$\frac{P_J *}{P_{mec} *} = \left(\frac{P_J}{M_m \Omega}\right)^* = \frac{J *}{r * \Omega} \qquad [5.31]$$

For constant speed, copper loss performance can be kept constant, regardless of the size if the current density varies with the scaling ratio. It thus becomes possible to design small motors with a good efficiency. This can be written:

$$\eta_J' = \frac{1}{1 + (1/\eta_J - 1)J * /(r * \Omega *)} \qquad [5.32]$$

5.2.6. *Example*

A reference motor with the following characteristics is considered:

– mechanical power: $P_{mec} = 1$ kW;

– rated speed: $\Omega_N = 3{,}000$ rpm;

– active diameter and length: ℓ,d = 100 mm;

– relative copper losses: $P_J/P_{mec} = 0.05$.

This last value corresponds to an efficiency of 95.2%.

The goal is to extrapolate the performances to a watch motor with the following characteristics:

– mechanical power: $P_{mec} = 1\mu W$;

– average speed: $\Omega_N = 30$ rpm.

In case of a reluctance or induction motor, relative values are obtained by [5.22]:

$$r^* = \left(\frac{P_{mec}^{\ *}}{\Omega^*} \right)^{1/4} \qquad [5.33]$$

with:

$$P_{mec}^{\ *} = 10^{-9}$$

$$\Omega^* = 10^{-2}$$

$$r^* = 1.78.10^{-2}$$

$$l' = d' = r^* l = 1.78 \text{ mm}$$

Using equation [5.25], efficiency is:

$\bullet_J' = 6.32\ 10^{-5} = 0.00632\%$!

For the case of a PM motor, by imposing a constant efficiency ($\bullet_J^*=1$), therefore also constant copper losses, the scaling factor is obtained from [5.30] and [5.32]:

$$r^* = \left(\frac{P_{mec}^{\ *}}{\Omega^{*2}} \right)^{1/5} = 0.1 \qquad [5.34]$$

The corresponding dimensions are too large to introduce such a motor into a watch. With an efficiency reduced to a value of 33.3% these values become:

$$\frac{P_J'}{P_{mec}'} = 2$$

$$\frac{P_J^{\ *}}{P_{mec}^{\ *}} = 40$$

Using relation [5.31], the scaling factor is:

$$r^* = \left[\frac{P_{mec}{}^*}{(P_J / P_{mec})^* \Omega^{*2}} \right]^{1/5} = 4.8\ 10^{-2} \qquad [5.35]$$

$\ell' \approx 4.8$ mm

In conclusion, a high efficiency for a small motor is only possible with an excitation by a permanent magnet structure. In this example, the power consumption with a magnet is 5,275 times weaker than without a magnet.

Other factors such as changing the relative copper or magnet volumes can improve the performances after reduction. Generally, a permanent magnet motor performs in a power range of μW to several kW. Beyond this, comparable performances can be achieved by induction or separate excitation motors.

5.3. Parametric expression

5.3.1. *Torque*

By using equations [5.22] and [5.30], the torque for a given motor can be expressed depending on its size. For a variable reluctance or induction motor:

$$M_{rel} = k_{rel} d^3 l \qquad [5.36]$$

For a length proportional to diameter:

$$M_{rel} = k_{rel}' d^4 \qquad [5.37]$$

For a PM motor:

$$M_{ai} = k_{ai} d^{2.5} l \qquad [5.38]$$

For a length proportional to diameter:

$$M_{ai} = k_{ai}' d^{3.5} \qquad [5.39]$$

Generally, for any motor with constant heating:

$$M = k_M d^n = k_M d^{n-1} l \qquad [5.40]$$

Knowledge of one motor of a series of a given type will determine the parameters *k* and thus express the other quantities.

5.3.2. *Comparison*

For the defined motor types (construction parameters such as magnet type, insulation class, powering mode, etc.), it is possible to draw the torque curves depending on the diameter for variable reluctance and PM motor types respectively (relations [5.37] and [5.39]). The curves of Figure 5.2 are thus obtained.

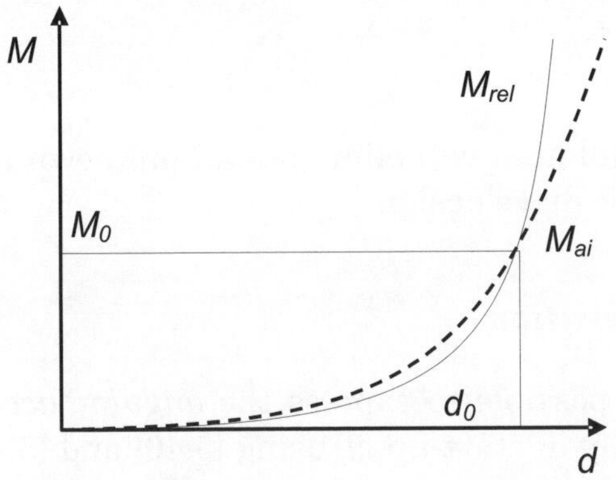

Figure 5.2. *Torque-diameter characteristics*

For small diameters, the PM motor torque is far greater than the induction or reluctance motor. For high diameters the opposite is the case. An intersection point of the two curves appears for a 100 to 150 mm bore diameter, or torques from 50 to 150 Nm.

5.3.3. *Inertia*

In particular, it is possible to express the moment of inertia of a cylindrical rotor.

$$J = \int R^2 dM = \frac{1}{2} R^2 m = \frac{1}{2} \rho \pi R^4 l = k_J d^4 l = k_J' d^5 \qquad [5.41]$$

With reference to relations [5.40] and [5.41], it is possible to express the inertia based on torque:

$$d = \left(\frac{M}{k_M} \right)^{1/n} \qquad [5.42]$$

$$J = k_J' \left(\frac{M}{k_M} \right)^{5/n} \qquad [5.43]$$

This formulation will eliminate an unknown in the search for an electric drive design.

5.3.4. *Acceleration*

It is also possible to express the *angular acceleration* for the motor only in start-up ε_d, using [5.40] and [5.41]:

$$\varepsilon_d = M / J = \frac{k_M}{k_J'} d^{n-5} \qquad [5.44]$$

According to the scaling laws, acceleration can be determined for both motor types:

$$\varepsilon_d{}^* = \frac{1}{r^*} \text{ for a reluctance or induction motor} \qquad [5.45]$$

$$\varepsilon_d{}^* = \frac{1}{r^{*1.5}} \text{ for a PM motor} \qquad [5.46]$$

Chapter 6

Global Design of an Electric Drive

6.1. Introduction

The choice of an electric drive cannot be dissociated from its devices:

– transmission, rotating or linear;

– power supply;

– control.

When a transfer time (travel time) is required for a given load, the motor choice and the corresponding transmission ratio constitute two related unknowns.

If the motor is initially known, it is possible to show that the optimum condition of the position transfer in a given time matches the following relation [3.44]:

$$r_{opt} = \sqrt{J_m / J_e} \qquad [6.1]$$

J_m = motor inertia;

J_e = driven body inertia.

If this relation is well known, it is not practical in reality. Indeed, for a given driven body, the motor is not known *a priori*. The choice of motor and transmission are free and optimized on multiple criteria. The following sections describe a general methodology approach.

6.2. Dynamic equations

6.2.1. *Position transfer*

A *position transfer* is generally combined with a *trapezium shape speed profile* (Figure 6.1), as defined in section 3.6. Acceleration and deceleration are realized at constant torque, corresponding to the motor and supply limits. The constant speed section corresponds in general to the motor speed limit. Relation [3.76] shows that the energy balance is optimal when the respective acceleration time (t_1), constant speed time ($T-t_1-t_2$) and braking time (t_2) are equal (Figure 6.1).

The angle α_e corresponds to the total movement (or x_e total linear movement) of the driven body. By integration, it has the following values:

$$\alpha_e = 2t_1\Omega_{me} = \frac{2}{3}T\Omega_{me} \text{ or } x_e = 2t_1 v_{me} = \frac{2}{3}Tv_{me} \qquad [6.2]$$

$$\Omega_{me} = \frac{3}{2}\frac{\alpha_e}{T} \text{ or } v_{me} = \frac{3}{2}\frac{x_e}{T} \qquad [6.3]$$

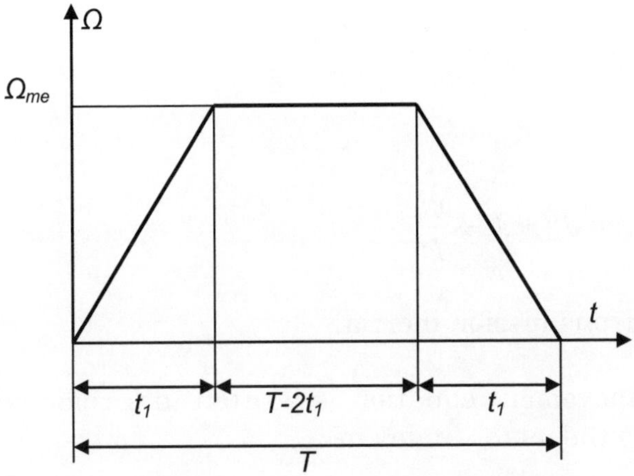

Figure 6.1. *Trapezium shape speed profile*

6.2.2. *Movement equation with a transmission*

The transmission ratio for a rotating (linear) system is defined as follows:

$$r = \frac{\Omega_m}{\Omega_e} \text{ [-] or } k = \frac{\Omega_m}{v_e} \text{ [rad/m]} \qquad [6.4]$$

Ω_m = motor speed: Ω_e, v_e = driven body speed;

J_e = driven inertia or: m_e = driven mass;

J_e' = driven inertia referred to the motor.

By kinetic energy conservation:

$$\frac{1}{2} J_e \Omega_e^2 = \frac{1}{2} J_e' \Omega_m^2 \qquad [6.5]$$

$$J_e' = \frac{J_e}{r^2} \text{ or } J_e' = \frac{m_e}{k^2} \qquad [6.6]$$

$$J_{tot} = J_m + J_r + \frac{J_e}{r^2}$$

or

$$J_{tot} = J_m + J_r + \frac{m_e}{k^2} \qquad [6.7]$$

J_r = transmission inertia.

The movement equation is written in a reference system linked to the motor (index m):

$$J_{tot} \frac{d\Omega_m}{dt} = M_m - M_f / r \qquad [6.8]$$

M_m = motor torque = M_o = rated torque;

M_f = friction torque of the driven body.

$$\frac{d\Omega_m}{dt} = r\varepsilon = r\frac{\Omega_{me}}{t_1} \quad \text{or} \quad \frac{d\Omega_m}{dt} = ka = k\frac{v_{me}}{t_1} \qquad [6.9]$$

The equations linking the main transfer parameters are, using [6.8] and [6.9]:

$$\left(J_m + J_r + \frac{J_e}{r^2}\right)\frac{r\Omega_{me}}{t_1} = M_0 - M_f / r \qquad [6.10]$$

or

$$\left(J_m + J_r + \frac{m_e}{k^2}\right)\frac{kv_{me}}{t_1} = M_0 - F_f / k \qquad [6.11]$$

The equation above establishes the correlation between the motor torque and the desired transfer time ($T = 3\ t_1$).

However, it contains an unwanted unknown: inertia J_m which depends on the torque M_0. The objective is to establish a correlation between torque and inertia by using expression [4.43].

6.2.3. *Solving*

The final equation [6.10] can be written in a form depending on the torque and gear ratio only, using [4.43]:

$$\left[k_J\left(\frac{M_0}{k_M}\right)^{5/n}+J_r+\frac{J_e}{r^2}\right]r\Omega_{me}=\left(M_0-\frac{M_f}{r}\right)t_1 \qquad [6.12]$$

For a rotating to linear transmission:

$$\left[k_J\left(\frac{M_0}{k_M}\right)^{5/n}+J_r+\frac{m_e}{k^2}\right]kv_{me}=\left(M_0-\frac{F_f}{k}\right)t_1 \qquad [6.13]$$

This relation establishes a correlation between the motor torque and the transmission ratio. From this, it is possible to take out the gear transmission expression as a torque function:

$$J_r+k_J\left(\frac{M_0}{k_M}\right)^{5/n}=J_{mr}\left(M_0\right)=J_{mr} \qquad [6.14]$$

$$J_{mr}\Omega_{me}r^2-M_0rt_1+J_e\Omega_{me}+M_ft_1=0 \qquad [6.15]$$

Two solutions are possible for the transmission ratio r:

$$r=\frac{M_0t_1\pm\sqrt{M_0^2t_1^2-4J_{mr}\Omega_{me}\left(J_e\Omega_{me}+M_ft_1\right)}}{2J_{mr}\Omega_{me}} \qquad [6.16]$$

A solution exists under the condition of a positive expression under the root. A limit value appears for a zero root:

$$M_0^2 t_1^2 - 4\left[J_r + k_J\left(\frac{M_0}{k_M}\right)^{5/n}\right]\Omega_{me}\left(J_e\Omega_{me} + M_f t_1\right) = 0 \qquad [6.17]$$

A limit value of M_o can satisfy the equation above, for M_{olim}, to which corresponds a limit transmission ratio:

$$r_{\lim} = \frac{M_{0\lim} t_1}{2J_{mr\lim}\Omega_{me}} \qquad [6.18]$$

For any value $M_o > M_{olim}$, a solution exists. The limit value is therefore the smallest possible motor.

For any other M_o, there are two values of the ratio r:

$$r_1 = \frac{M_0 t_1 + \sqrt{\ }}{2J_{mr}\Omega_{me}} \text{ and } r_2 = \frac{M_0 t_1 - \sqrt{\ }}{2J_{mr}\Omega_{me}} \qquad [6.19]$$

Only the value r_2 presents an interest. Indeed, the kinetic energy in the start-up is always lower for the ratio r_2:

$$W_{kin}(r_2) < W_{kin}(r_1)$$

The kinetic energy in start-up is written:

$$W_{kin} = \frac{1}{2} J_{tot}\Omega_{me}^2 = \frac{1}{2}\left[k_j\left(\frac{M_0}{k_m}\right)^{5/n} + J_r + \frac{J_e}{r^2}\right] r^2\Omega_{me}^2 \qquad [6.20]$$

6.3. Example

6.3.1. *Data*

The chosen example is borrowed from a linear drive application for a machine to punch laminations. The system is associated with two X-Y axes, moved by two rotating motors associated with ball screws. The specification is the following:

– Driven mass $m_e = 30$ kg;

– Rated travel $x_e = 30$ mm;

– Transfer period $T_{tot} = 0.375$ s;

– Transfer time $t_1 = 0.2$ s;

– Idle time $t_0 = 0.175$ s;

– Friction force $F_f = 30$ N;

– Position accuracy ± 0.25 mm.

6.3.2. *Drive chosen by the manufacturer*

The machine manufacturer completed the following choice, corresponding to an existing DC-motor:

– Torque constant $k_M = 0.353$ Nm/A;

– Maximum speed 4,000 rpm;

– Rotor resistance $R_i = 0.95$ Ω;

– Rated torque $M_N = 3$ Nm;

– Maximum torque $M_{max} = 15$ Nm;

– Rated current $I_N = 8.5$ A

– Rotor inertia $J_m = 1.5.10^{-3}$ kgm²;

– Thermal resistance $R_{th} = 1.3$ K/W;

– Ball screw pitch = 4 mm;

– Screw + coupling inertia J_v = 171 10^{-6} kgm^2 (referred to the motor).

The first two motors mounted on this machine were destroyed in a couple of minutes. Knowing the manufacturer had chosen an acceleration and deceleration at maximum torque mode M_{max}, the first objective is to determine the causes of failure.

The rotating-linear transmission ratio is:

$$k = 2\pi / step = 1{,}571 \text{ m}^{-1}$$

The total rotor inertia including coupling and screw is:

$$J_{totm} = 1.671.10^{-3} \text{ kgm}^2$$

As a result, the corresponding mass reported [3.13] at the driven body becomes:

$$m'_{totm} = k^2 J_{totm} = 4{,}124 \text{ kg}$$

This value is 137 times higher than that of the moving lamination mass. The role of a motor is primarily to accelerate itself.

The force in acceleration functioning has the expression:

$$F_m = M_{\max} k = 2.36.10^4 \text{ N}$$

The acceleration a has the expression:

$$a = F_m /(m'_{totm} + m_e) = 5.67 \text{ m/s}^2$$

Referring to Figure 6.2, the moving x_e becomes:

$$x_e = a t_a (t_1 - t_a) \qquad [6.21]$$

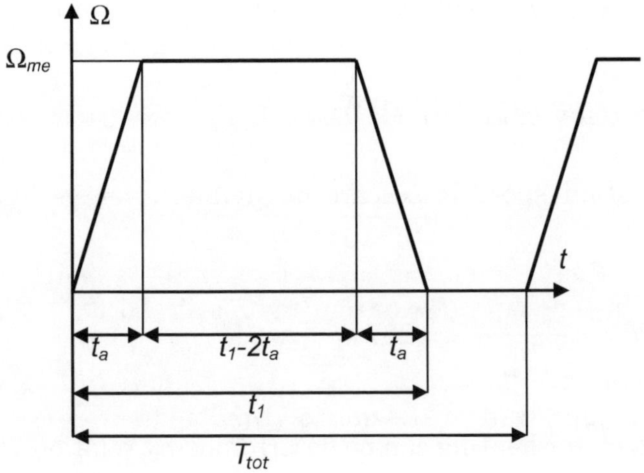

Figure 6.2. *Trapezium-shaped speed profile with maximum acceleration*

The acceleration time t_a is then:

$$t_a = \frac{at_1 - \sqrt{a^2 t_1^2 - 4ax_e}}{2a} \qquad [6.22]$$

$t_a = 0.0314$ s

The corresponding speed is:

$v_e = at_a = 0.178$ m/s

$\Omega_m = kv_e = 280$ rad/s

$N_m = 2{,}670$ Rpm

On this basis, the copper losses can be determined, in acceleration mode:

$$I_{\max} = 5I_N = 42.5 \text{ A}$$

$$P_{J\max} = RI_{\max}^2 = 1{,}716 \text{ W}$$

Constant speed losses are negligible. Average losses are then:

$$\overline{P}_J = \frac{2t_a P_{J\max}}{T_{tot}} = 287 \text{ W} \qquad [6.23]$$

The motor heating can be determined by relation [7.6]:

$$\Delta T = R_{th} \overline{P}_J = 374 \text{ K}$$

With such motor supporting a maximum temperature of 130 °C, it is therefore normal that it quickly slams.

6.3.3. *New drive with the same motor type*

In order to minimize the losses, the acceleration must be imposed during 1/3 period [3.76], the same applies for the deceleration. It is therefore: ta = 0.0667 s. From [6.21], acceleration is:

$$a = \frac{x_e}{t_a(t_1 - t_a)} = \frac{9x_e}{2t_1^2} = 3.375 \text{ m/s}^2 \qquad [6.24]$$

$$v_{\max} = at_a = 0.225 \text{ m/s}$$

To reduce the motor torque and the corresponding inertia, it is necessary to choose as high a screw pitch as possible, yet compatible with the accuracy required and corresponding normalized values. A 12 mm pitch seems a good compromise. In this case, for a precision of ± 0.25 mm, the motor resolution is 48 steps per revolution, which is easy to fulfill.

The result is the following for the transmission ratio:

$$k = 2\pi \,/\, step = 523.6 \text{ m}^{-1}$$

The screw and coupling inertia with respect to load is:

$$m'_{v+a} = k^2\, 161.10^{-6} = 47 \text{ kg}$$

The current in acceleration regime I_a can be linked to the rated current by [6.23]:

$$\overline{P}_{JN} = RI_N^2 = RI_a^2 \frac{2t_a}{T_{tot}} \qquad [6.25]$$

It infers the relationship between the acceleration current and the rated current:

$$I_a = \sqrt{\frac{T_{tot}}{2t_a}} I_N = 1.68\, I_N \qquad [6.26]$$

Motor torque and current are proportional for a PM motor:

$$M_a = k_M I_a = 1.68 k_M I_N = 1.68 M_N$$

By choosing a new motor with the same manufacturing as that which was chosen by the punching machine designer, the initial rotor inertia and motor torque (index i) and new values (without index) are linked by [5.43]:

$$J_m = J_{mi}\left(\frac{M_N}{M_{Ni}}\right)^{10/7} \qquad [6.27]$$

By replacing in the expression of acceleration [6.24], we obtain:

$$a = \frac{kM_N\sqrt{T_{tot}/2t_a}}{m_e + k^2\left[J_{v+a} + J_{mi}\left(M_N/M_{Ni}\right)^{10/7}\right]} \qquad [6.28]$$

By replacing with numerical values, the corresponding expressions are:

$$0.295 + 0.3825\, M_N^{10/7} = M_N$$

This equation has two solutions:

M_N = 0.38 Nm

M_N = 13 Nm

Of course the first solution is retained, and the following quantities correspond to it:

M_a = 1.68 M_N = 0.635 Nm;

J_m = 7.84.10-5 kgm^2;

m'_m = rotor equivalent mass = 21.5 kg

N_{max} = maximum speed = 1,125 rpm

A parametric analysis with screw pitch from 4 to 20 mm leads to Table 6.1.

Pitch [mm]	Transmission Ratio [m^{-1}]	m_{v+a}' [kg]	M_N [mNm]	M_a [Nm]	J_m [$kgm^2.10^{-6}$]	m_m' [kg]	N_{max} [Rpm]	Resolution [step/rev]
4	1571	397	-	-	-	-	3380	16
8	785.4	99	527	885	125	77	1690	32
12	523.6	47	380	635	78	21.5	1125	48
16	392.7	25	330	554	64	9.9	845	64
20	314.1	16	334	561	65	6.5	675	80

Table 6.1. *Motor and transmission parametric analysis for various screw pitches*

For a 4 mm pitch, a motor of the type considered can certainly achieve the required performance. Figure 6.3 shows the evolution of the motor rated torque and reported equivalent mass of inertias (motor + screw and coupling) as a function of the screw pitch.

In conclusion, the electric drive chosen exhibited two errors:

– an unnecessarily high transmission ratio, with a pitch of 4 mm, it is impossible to find a motor with the previously chosen quality that satisfies thermal criteria;

– an overly powerful motor, leading to an inertia such as the motor has for a main task to accelerate itself.

However, for a lower transmission ratio (pitch from 8 to 20 mm), there is no problem finding a solution, with a much less powerful motor.

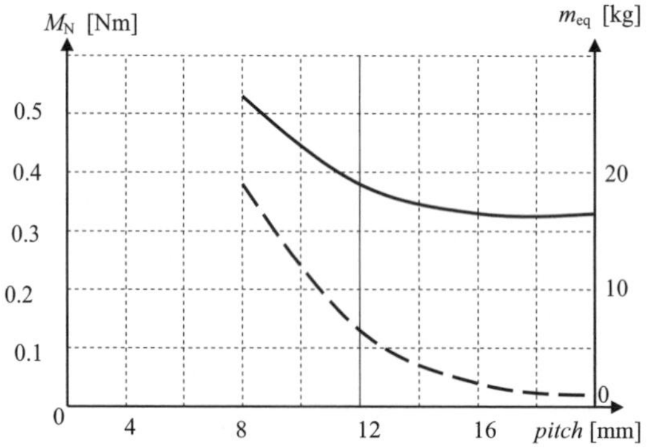

Figure 6.3. *Motor rated torque and equivalent mass of motor + screw + coupling as a function of the screw pitch*

6.3.4. *Drive design with a motor of a given type*

Analysis of the preceding section has been based on an existing implementation fixing as its improvement is an objective.

In practice, the approach is to start from specification and a prior choice of a motor type.

The choice is a synchronous self-commutated motor with permanent magnets and high-performance mass torque. The following values are drawn from a catalog:

$Mo = k_M\, d^{3.5}$;

$k_{M1} = 1{\times}10^5$ high performance motor;

$k_{M2} = 5{\times}10^4$ average performance motor.

Initially, the transmission inertia has been neglected. The rotor inertia coefficient is valid for both cases:

$J = k_J\, d^5$;

$k_J = 768$.

In an initial approximation, the inertia of the screw and the coupling is assumed to be double that of the rotor.

6.3.5. *High performance motor*

For this motor, we obtain the following limit values [6.17], [6.18]:

M_{olim} = 74.33 mNm;

k_{lim} = 2,720 corresponding to a ball screw pitch of 2.33 mm;

d_{lim} = 17.73 mm;

N_{lim} = 5,800 rpm;

W_{cin} = 1.52 J.

This limit solution is not acceptable, due to non-integer screw pitch (in mm) and its limited character (sensitivity to changes in external parameters such as friction).

By varying the screw pitch from 4 to 16 mm, the results in Table 6.2 are obtained.

Pitch [mm]	**4**	**6**	**8**	**10**	**12**	**16**
M_0 [mNm]	95	127.5	162.5	195	230	298
N_0 [rpm]	3360	2251	1672	1351	1124	840
d [mm]	19	20.7	22.2	23.4	24.5	26.4
W_{kin} [J]	1.12	1.00	0.949	0.920	0.899	0.873

Table 6.2. *Performance evolution based on the screw pitch for a high performance motor*

Figure 6.4 shows torque, speed, and kinetic energy related to the screw pitch.

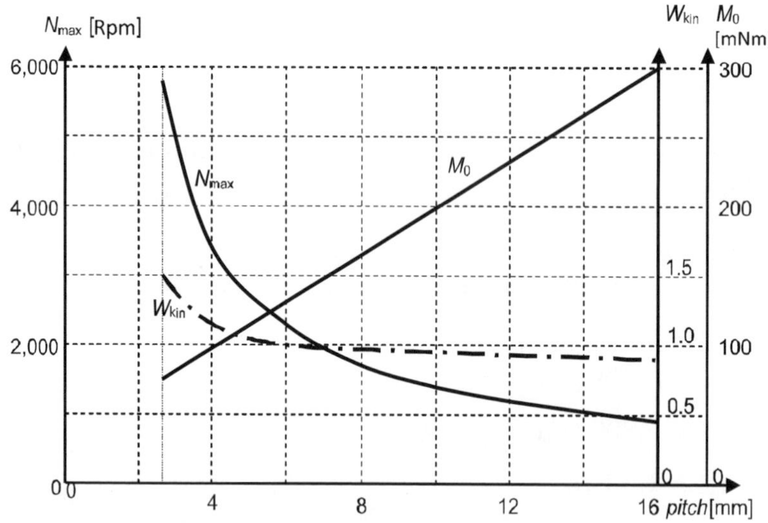

Figure 6.4. *Changes in the motor rated torque, maximum speed and kinetic energy related to the screw pitch*

6.3.6. *Average performance motor*

For this motor, we obtain the following limit values [6.17], [6.18]:

$M_{olim} = 420.5$ mNm;

$k_{lim} = 481$ corresponding to a ball screw pitch of 13.1 mm;

$d_{lim} = 35.4$ mm;

$N_{lim} = 1{,}028$ rpm;

$W_{kin} = 1.52$ J.

This limit solution is also not acceptable, due to non-integer screw pitch (in mm) and its limited character (sensitivity to changes in external parameters such as friction).

By varying the screw pitch from 15 to 20 mm, the results in Table 6.3 are obtained.

Pitch [mm]	**15**	**16**	**18**	**20**
M_0 [mNm]	431	442	468	496
N_0 [Rpm]	900	840	748	675
d [mm]	35.7	36.0	36.8	37.2
W_{kin} [J]	1.35	1.30	1.22	1.17

Table 6.3. *Performance evolution based on the screw pitch for an average performance motor*

6.4. Conclusions

The advantage of the methodology described is an analysis without *a priori* knowledge of the optimal solution for the motor and transmission choice. If there is usually no optimum, there is however a more favorable solution taking account of constraints such as speed, resolution, screw pitch, transmission limit, energy balance, etc. The approach illustrates the importance of motor performance (factor k_M), linked to the structure type.

This methodology should of course still be compared with catalogs, which limit solutions.

Chapter 7

Heating and Thermal Limits

7.1. Heating importance

The heating is the main limiting factor of the permissible power for a motor. The heating is defined by losses, the motor structure and the cooling mode: convection, external fan, internal fan, air-air heat exchanger, water-air exchanger, direct cooling by water or oil, etc. It is the insulation thermal quality which defines a maximum temperature limit. The study of the thermal behavior of electric machines is necessary for defining the transient evolution and acceptable temporary overloads.

The thermal behavior is primarily based on three phenomena:

– conduction;

– convection;

– radiation.

7.2. Thermal equations

7.2.1. *Conduction*

For a medium of thickness e_1 and of surface S_1, the conduction is governed by the following dynamic equation:

$$P = cV\frac{\partial T}{\partial t} + \frac{\lambda S_1}{e_1}(T - T_e) \qquad [7.1]$$

P = system losses [W];

V = volume concerned with heat accumulation volume [m^3];

T = temperature [K];

λ = thermal conductivity [W/mK];

T_e = external temperature [K];

c = specific heat per volume [J/m^{3}K].

This equation is characteristic of the internal conduction phenomenon in an electrical machine. The stabilized solution of this equation, corresponding to the steady state and to the particular solution is:

$$T_1 - T_e = P\frac{e_1}{\lambda S_1} \qquad [7.2]$$

7.2.2. *Convection and radiation*

Convection and radiation are characteristic phenomena of thermal losses to ambient air evacuation. A coefficient for these two phenomena can be defined, characterized as follows:

α = equivalent convection coefficient $\approx 12+6v^{0.8}$ [W/m²K] [7.3]

with v = air speed [m/s].

The characteristic convection equation is written:

$$P = cV\frac{\partial T}{\partial t} + \alpha S_2\left(T - T_e\right) \quad [7.4]$$

with S_2 = convection surface.

This equation is characteristic of convection phenomenon for an electric machine surface. This equation stabilized solution corresponds to the steady state and to the particular solution:

$$T_2 - T_e = P\frac{1}{\alpha S_2} \quad [7.5]$$

7.2.3. *Global phenomenon*

Conduction and convection appear simultaneously in an electric motor. It is possible to reduce the phenomenon as a whole to an equivalent convection. In an initial step, the conduction and convection can be considered as two conduction phenomena in series. The *thermal resistance* can be written, in steady state:

$$R_{th} = \frac{\Delta T}{P} \quad [7.6]$$

ΔT = thermal difference = $T - T_e$.

As losses are evacuated through the two conduction and convection resistances, they are respectively:

$$R_{th1} = \frac{e}{\lambda S_1} \text{ (for conduction)} \qquad [7.7]$$

$$R_{th2} = \frac{1}{\alpha S_2} = \frac{e_2}{\lambda S_2} \text{ (for convection)} \qquad [7.8]$$

with:

$$e_2 = \lambda / \alpha$$

The total thermal resistance can be written for conduction and convection elements in series:

$$R_{th} = R_{th1} + R_{th2} = \frac{e}{\lambda S_1} + \frac{1}{\alpha S_2} \qquad [7.9]$$

The phenomenon can be reduced to a resulting convection associated with the S_2 surface;

$$R_{theq} = \frac{1}{\alpha_{eq} S_2} = \frac{e}{\lambda S_1} + \frac{1}{\alpha S_2} \qquad [7.10]$$

$$\alpha_{eq} = \frac{1}{\lambda S_1 + 1/\alpha} \qquad [7.11]$$

The full transient equation for two phenomena thus becomes:

$$P = cV \frac{\partial T}{\partial t} + \alpha_{eq} S_2 (T - T_e) \qquad [7.12]$$

or in a form enabling the resolution:

$$\frac{\partial T}{\partial t} + \frac{\alpha_{eq} S_2}{cV} T = \frac{P + T_e \alpha_{eq} S_2}{cV} \qquad [7.13]$$

7.2.4. *Resolution*

The resolution of the resulting heat equation [7.13] allows us to determine the following main quantities (Figure 7.1):

– the *stabilized temperature* (T_S, corresponding to the particular solution):

$$T_s = T_e + \frac{P}{\alpha_{eq} S_2} = T_e + PR_{theq} \qquad [7.14]$$

– the thermal time constant:

$$\tau_{th} = \frac{cV}{\alpha_{eq} S_2} = cVR_{theq} \qquad [7.15]$$

The transient solution is:

$$T = T_e + (T_s - T_e)\left(1 - e^{-t/\tau th}\right) \qquad [7.16]$$

We can define, by analogy with the electric field, an equivalent thermal capacity C_{theq}:

$$C_{theq} = cV \qquad [7.17]$$

$$\tau_{th} = R_{theq} C_{theq} \qquad [7.18]$$

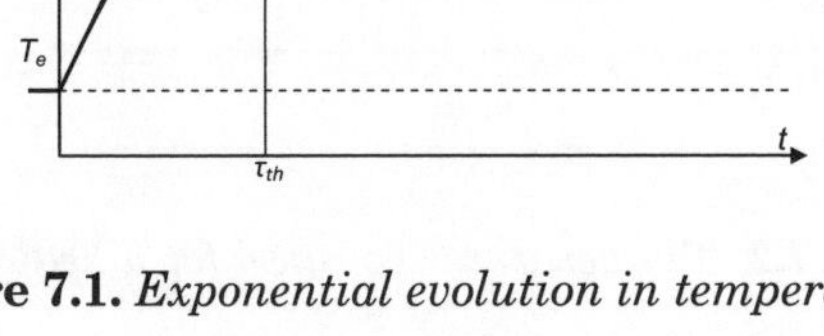

Figure 7.1. *Exponential evolution in temperature*

7.2.5. *Measurement*

For a motor, there are generally no known τ_{th} and T_s quantities for a given load. To determine them, a test support can be done for which the losses (additional test) are known. The measurement of the thermal evolution enables us to easily determine the time constant τ_{th} and the proportionality ratio between losses and stabilized heating.

$$\frac{T_s - T_e}{P} = \frac{1}{\alpha_{eq} S} = R_{theq} \qquad [7.19]$$

This coefficient is constant for a constant speed.

7.2.6. *Start-up*

Induction, or DC motor start-up operation, generates a peak current and the accumulation of a certain amount of thermal energy, mainly in the rotor. This operation can be linked to a specific thermal constant. Stabilized temperature is not determined practically, taking into account the speed evolution (Figure 7.2).

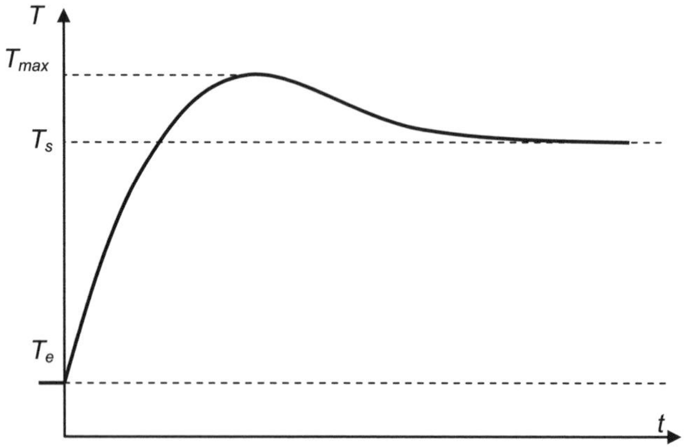

Figure 7.2. *Temperature evolution for a heavy start-up*

7.2.7. *Variable behavior*

Figure 7.1 characterizes a thermal behavior linked to variation in load or electrical conditions. Knowledge of any power and load transient conditions may be reduced to a succession of exponential evolutions.

For example, it is thus possible to determine the maximum peak temperature or the permissible period of temporary overload. The following figures describe several examples of the thermal evolution based on load.

Figure 7.3 describes the temperature evolution for a succession of identical cycles shorter than the transient thermal time constant. Figure 7.4 describes the temperature evolution to a succession of identical cycles at steady state.

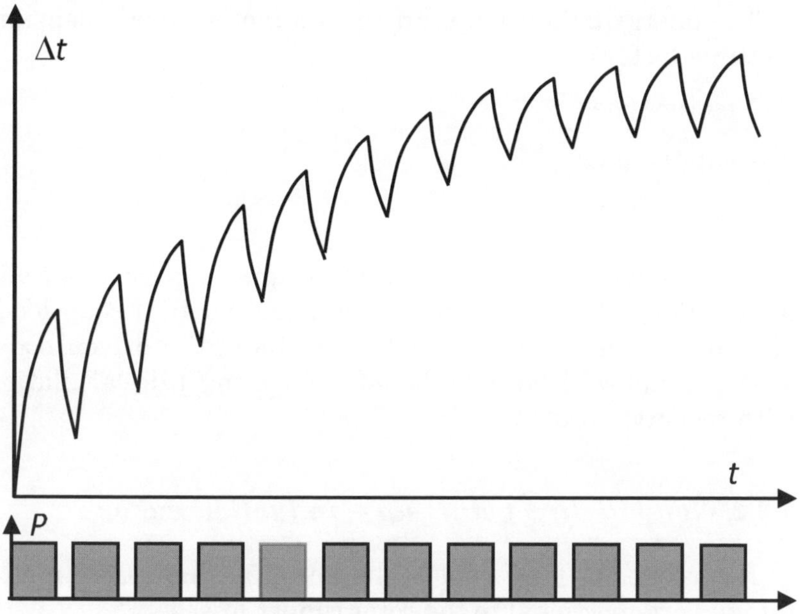

Figure 7.3. *Temperature evolution in a succession of identical transient cycles*

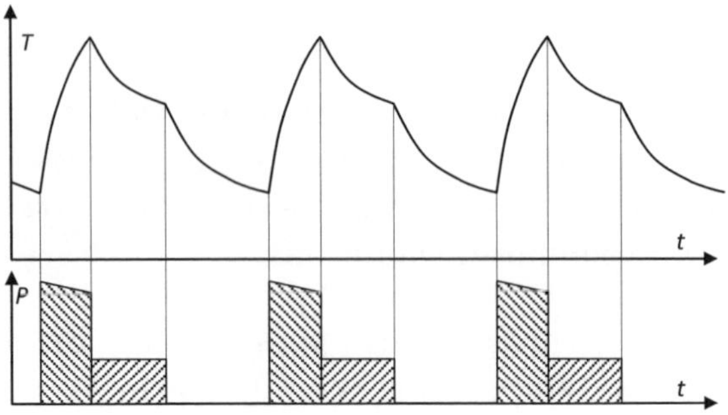

Figure 7.4. *Temperature evolution in a succession of identical steady cycles*

7.3. Energy dissipated at start-up

7.3.1. *Start-up conditions*

The energy balance at start-up of a motor mainly depends on three factors:

– motor type,

– supply mode,

– load.

With regard to load, resistant torque is a very specific factor. Very often a motor is characterized by its no load-start-up, taking into account the inertia of the driven body. This is what will be done below, indicating full calculation with resistive torque.

7.3.2. *No-load direct start-up – induction motor*

Referring to [JUF 95a, Chapter 15], induction motor torque is proportional to the rotor copper losses:

$$M_{as} = \frac{P_{Jr}}{s\Omega_s} \qquad [7.20]$$

with:

$$s = \frac{\Omega_s - \Omega}{\Omega_s} = \text{slip} \qquad [7.21]$$

Ω_s = synchronous speed

Furthermore, the movement equation is written:

$$J_{tot}\frac{\partial \Omega}{\partial t} = M_{as} \qquad [7.22]$$

Start-up energy spent in rotor copper is thus, using [7.20], [7.21] and [7.22]:

$$W_{thas} = \int P_{jr}dt = \int J_{tot}\frac{\partial \Omega}{\partial t}s\Omega_s dt$$

$$W_{thas} = \int_0^{\Omega s} J_{tot}(\Omega_s - \Omega)d\Omega = \frac{1}{2}J_{tot}\Omega_s^2 \qquad [7.23]$$

The energy dissipated in the rotor copper is equal to the group kinetic energy. In the case of a load with high resistant torque:

$$W_{thas} \geq W_{kin} \qquad [7.24]$$

This assessment does not take into account stator copper losses, which are of a comparable order of magnitude.

7.3.3. *No-load direct start up – DC motor*

Referring to [JUF 95a] and [4.8], a DC motor torque at constant excitation is written:

$$M = k_M I \qquad [7.25]$$

The current is then:

$$I = \frac{U - U_i}{R} = \frac{U - k_M \Omega}{R} \qquad [7.26]$$

k_M = torque or back-EMF (U_i) coefficient

Energy dissipated to the rotor to start-up is, using [7.22], [7.25] and [7.26]:

$$W_{thcc} = \int RI.Idt = \int (U - k_M \Omega) \frac{J}{k_M} d\Omega$$

with $\Omega_0 = \frac{U}{k_M}$

$$W_{thcc} = \int_0^{\Omega o} (\Omega_0 - \Omega) J d\Omega = \frac{1}{2} J \Omega_0^2 = W_{kin} \qquad [7.27]$$

Energy dissipated in the rotor is equal to the group kinetic energy. In the case of a load with an important torque:

$$W_{thcc} \geq W_{kin} \qquad [7.28]$$

This assessment does not take into account copper losses to the stator, which are significantly lower (separate excitation or shunt), even zero in the case of permanent magnets.

7.3.4. *Variable frequency start-up – induction motor*

With reference to section 4.3.5, the torque of an induction motor powered by a variable frequency is proportional to the relative speed difference between the rotor and stator rotating field, in the linear torque characteristic area. By imposing a constant torque, it is possible to write for the slip and rotor copper losses:

$$s = \frac{\Delta\Omega}{\Omega_s} \text{ with } \Delta\Omega = \text{constant}$$

$$W_{Jr} = \int s M \Omega_s dt \qquad [7.29]$$

By replacing the torque M with [7.22] valid at no load, the expression becomes:

$$W_{Jas} = J\Delta\Omega \int_0^{\Omega s} d\Omega = J\Omega_s \Delta\Omega \qquad [7.30]$$

Referring to the rated speed (at the rated frequency) for which:

$$s = s_N \text{ and } \Omega_s = \Omega_0$$

$$W_{Jr} = J\Omega_0^2 s_N = 2 s_N W_{kin} \qquad [7.31]$$

This balance is much more favorable than a direct start-up [7.27], in a ratio of $2s_N$ or 0.4 for a small motor and down to 0.04 for a large one.

7.3.5. *Variable voltage start-up – DC motor*

With reference to [7.22], [7.23] and [7.26], a variable voltage start-up imposing a constant torque, thus a constant current I_N, leads to the following expression:

$$RI_N = \gamma U_N$$

$$U_N = k_M \Omega_0 \qquad [7.32]$$

$$W_{Jcc} = \int \gamma U_N I_N dt = \int \gamma U_N \frac{Jd\Omega}{k_M} = \int_0^{\Omega_0} \gamma \Omega_0 Jd\Omega = \gamma J \Omega_0^2 = 2\gamma W_{kin}$$

[7.33]

This balance is much more favorable than a direct start-up [7.27], in a ratio of 2γ or 0.2 for a small motor and down to 0.02 for a large one.

7.3.6. *Brushless DC motor start-up*

Similar to the DC motor supplied by variable voltage, Figure 3.6 characterizes the start-up behavior, with constant current until the steady speed Ω_0. Relations [7.32] and [7.33] remain valid.

7.4. Cooling modes

7.4.1. *Techniques used*

Motor cooling may be provided by various processes, depending on the motor size, mainly using conduction and convection. The following processes are used:

– natural conduction and convection for small motors;

– forced conduction and convection by external circulation of air around the housing for a few kW motors;

– forced conduction and convection by internal circulation of air for motors from several kW to hundreds of kW;

– forced conduction and convection, internal and external, by air-to-air heat exchanger for motors of several hundred kW, in dirty atmosphere;

– forced conduction and convection using air-to-water heat exchanger for motors from hundreds of kW to MW, in dirty atmospheres or for an efficient cooling;

– conduction and convection by direct water cooling in the stator, for motors of several MW, for very efficient motor cooling.

7.4.2. *Air cooling*

For even higher powers, of the order of tens of kW and beyond, a fan mounted on the shaft provides internal forced air circulation in the motor (Figure 7.6). In the latter case, if the speed is highly variable, especially with frequent low-speed operation, air forced by a fan driven by a separate motor is necessary. Figure 7.7 shows two examples of this solution.

When ambient air is polluted (dust, moisture or aggressive atmosphere), an air-air tube heat exchanger is used in opposite-flow or cross-flow. Figure 7.8a presents an example of tubes placed around the housing and Figure 7.8b describes a solution for an exchanger placed above the motor.

Figure 7.5. *External fan-cooled motor with housing fins*

Figure 7.6. *Motor cooled by internal fan and air circulation*

Figure 7.7. *Internal convection with separate motor-fan cooled motors*

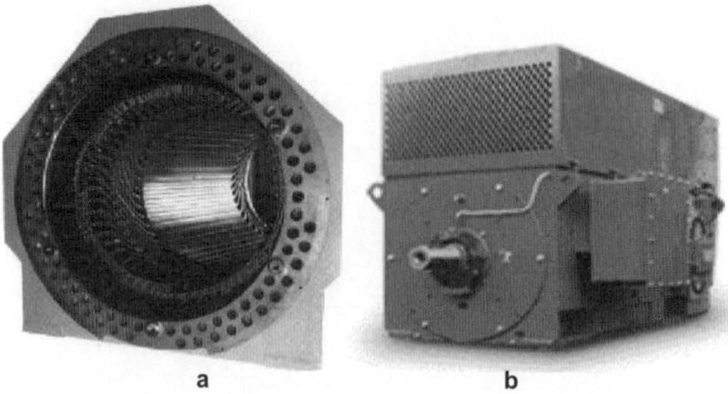

a b

Figure 7.8. *Air-air exchanger for motors cooled by internal convection: a) tubes placed around the housing; b) tubes placed above the motor*

7.4.3. *Water cooling*

For some high powers of several MW, cooling water is sometimes used around the housing or even in the winding bars for very high powers, indirectly by a water-air exchanger or directly by water circulation introduced into the stator.

Figure 7.9 shows a water-air exchanger placed on top of the motor. Figure 7.10 shows the drawing of a 45 MW motor-generator with separate cooling imposed by the dual rotating direction of a pump-turbine group. Four groups with water-air exchangers and fans are placed on the outskirts of the stator.

Very large motors or alternators are sometimes directly water-cooled, in particular by water circulation in the stator bars. Figure 7.11 shows water supply tubes in the stator. Figure 7.12 shows the water tube distribution, arranged between partial conductors.

Figure 7.9. *Water cooled motor, with internal convection and water-air exchanger*

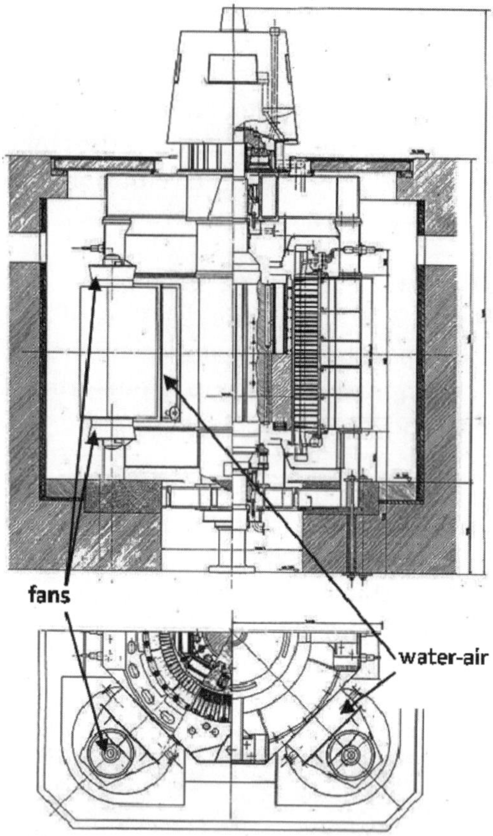

Figure 7.10. *Motor-generator with separate cooling by four groups with water-air exchangers and fans*

Figure 7.11. *Motor-generator with water supply by tubes in the stator bars*

Figure 7.12. *Roebel's bars with water tube distribution, arranged between partial conductors*

Chapter 8

Electrical Peripherals

8.1. Adaptation

Any electrical drive can include many *electrical peripheral* elements delivering adaptation functions between the source and electric drive. Three main categories can be distinguished:

– electrical or electromagnetic adaptation systems;

– power electronic or control adaptation systems;

– the sensors enabling control and regulation.

This chapter deals with electrical and electromagnetic peripherals only.

8.2. Sources

All the usual sources are real voltage sources, characterized by an ideal voltage source and an internal resistance or impedance (Figure 8.1). The internal impedance can be obtained by a brief duration short-circuit. For an alternative network, internal impedance is generally inductive (power line), but can be capacitive (cable). DC voltage sources, resulting from the conversion of AC (by rectifier) are generally resistive with an inductive component that may play a role in the case of transient behavior. DC voltage sources of a chemical nature are instead of a capacitive nature. They are generally characterized by a strong dependence on thermal internal resistance: this increases strongly as the temperature decreases.

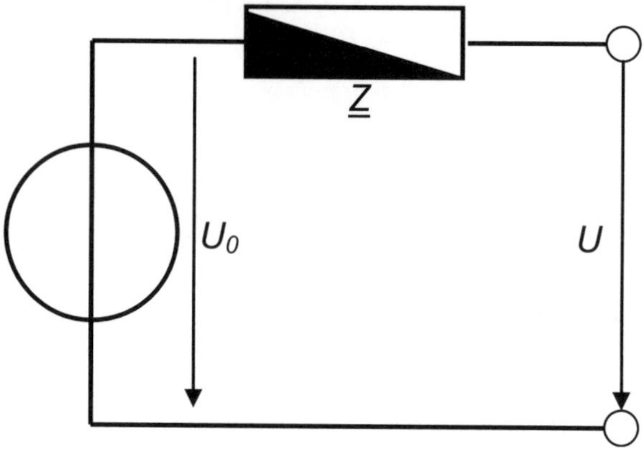

Figure 8.1. *Real voltage source*

8.3. Voltage adjustment

8.3.1. *Principle*

Voltage adaptation also implies current adaptation. It is generally characterized by a constant voltage adaptation at the load terminals.

Adaptation of AC voltage is obviously achievable by means of a transformer. The reasons for this adaptation are related to security, power or economic criteria.

This adaptation may be made temporarily, for example for starting an induction motor by limiting the inrush current. In this case it will frequently involve an autotransformer.

8.3.2. *Autotransformer*

An *autotransformer* includes a single winding separated into two parts as shown in Figure 8.2. Primary and secondary have a common point. The characteristic equations are, for an ideal transformer:

$$\ddot{u} = \frac{U_1}{U_2} = \frac{I_2}{I_1} = \text{transformation ratio} \qquad [8.1]$$

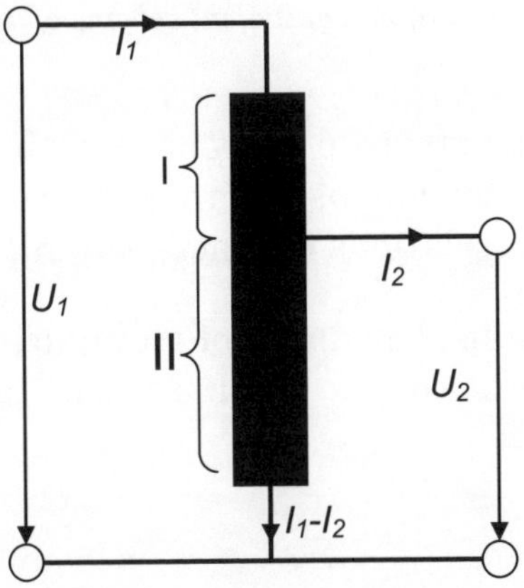

Figure 8.2. *Autotransformer*

The apparent powers of winding parts I and II are respectively:

$$S_I = (U_1 - U_2)I_1 = U_1 I_1 (1 - 1/\ddot{u}) \qquad [8.2]$$

$$S_{II} = |(I_1 - I_2)|U_2 = U_2 I_2 (1 - 1/\ddot{u}) \qquad [8.3]$$

The apparent power of an autotransformer is reduced by a factor (1 - 1/ü). This is the main advantage of this technology. Generally, a transformer or an autotransformer can be schematized by an ideal transformer and an impedance in series (Figure 8.3).

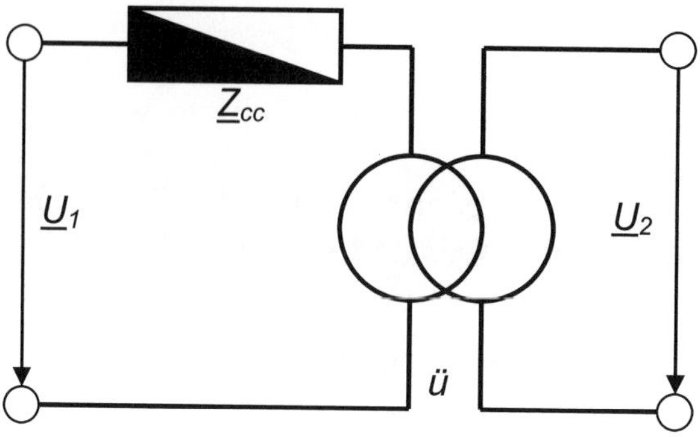

Figure 8.3. *Simplified equivalent scheme of a transformer*

By reducing the start-up current of an induction motor of a $\underline{Z}_M$ impedance by means of an autotransformer (Figure 8.4), the following relations characterize the behavior:

$\underline{Z}_M$ is the start-up impedance of an induction motor at zero speed. The impedance $\underline{Z}'_M$ with respect to the primary is then:

$$\underline{Z}'_M = ü^2 \underline{Z}_M \qquad [8.4]$$

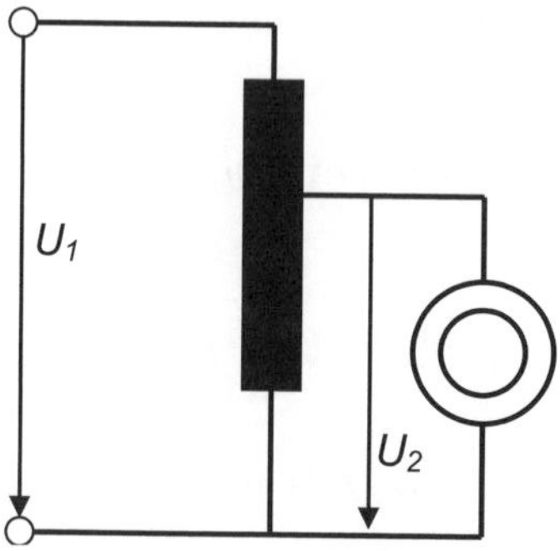

Figure 8.4. *Induction motor start-up using an autotransformer*

If the primary start-up current is imposed on a I_{1d} limit value, the impedance becomes:

$$\left|\ddot{u}^2 \underline{Z}_M + \underline{Z}_{cc1}\right| = \frac{U_1}{I_1} \qquad [8.5]$$

It is thus possible to infer the transformation ratio value. In the particular case for which the transformer impedance is negligible, we obtain:

$$\ddot{u} = \sqrt{\frac{U_1}{I_1 \underline{Z}_M}} \qquad [8.6]$$

For a given speed, the induction motor torque is proportional to the voltage square:

$$M_M \approx U_M^2 \qquad [8.7]$$

Always for a given speed, the torque with autotransformer M_{Mat} added to the full voltage torque M_M is:

$$\frac{M_{Mat}}{M_M} = \left(\frac{U_M}{U_1}\right)^2 = \frac{1}{\ddot{u}^2} \qquad [8.8]$$

For direct start-up, the current I_{Md} is:

$$I_{Md} = \frac{U_1}{Z_M} \qquad [8.9]$$

With an autotransformer, the start-up current is:

$$I_{1at} = \frac{U_1}{\ddot{u}^2 Z_M} = \frac{I_{Md}}{\ddot{u}^2} \qquad [8.10]$$

It therefore has a reduction in the same proportions of torque [8.8] and primary current [8.10].

Start-up with an autotransformer involves a switching operation from secondary to primary then to full voltage at the end of the start-up. In this case, "Korndörfer" coupling is used, characterized by opening the common point (Figure 8.5).

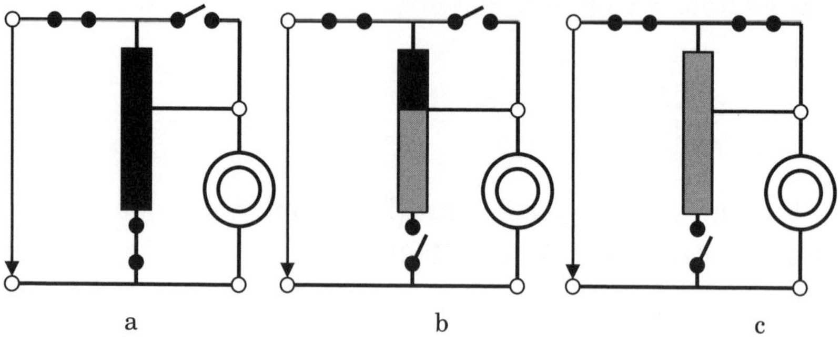

Figure 8.5. *Korndörfer start-up principle*

Switching from reduced voltage to full voltage cannot be done directly. Indeed, such direct switching corresponds to the short-circuit of a winding part of the autotransformer and its destruction. The switching process described in Figure 8.5, with a transition state (Figure 8.5b) introducing an inductance in series which is then short-circuited (Figure 8.5.c), avoids this problem.

8.3.3. *Star-triangle start-up*

To start induction motors from the order of a few kW to 100 kW, a reduction of the current can be achieved by a change of coupling. To do this, the motor will be supplied in triangle in normal mode and started in star scheme. It follows the following behavior:

$$\begin{cases} U_{ph\lambda} = \dfrac{U_{ph\Delta}}{\sqrt{3}} \\ I_{1\lambda} = \dfrac{I_{1\Delta}}{3} \end{cases} \qquad [8.11]$$

$$M_{\lambda} = \frac{M_{\Delta}}{3}$$

Even though this reduction principle is simple, it nevertheless has three drawbacks:

– the reduction of current and torque is necessarily 3;

– the circuit breaker switch $\lambda - \Delta$ is a relatively expensive device for high power;

– the peak of current switching $\lambda - \Delta$ is short, but very high.

8.4. Current adjustment devices

8.4.1. *Principle*

There is no fundamental difference between a voltage or current reduction associated with a motor. Current adaptation may occur mainly by two different methods:

– adaptation by a controlled current source. This approach will be analyzed in Chapter 9;

– adaptation by placing a linear component in series (R, L or C). This solution will be analyzed in the following sections.

8.4.2. *Start-up resistance*

Starting a DC motor with a current reduction can be performed by putting a resistance in series (Figure 8.6).

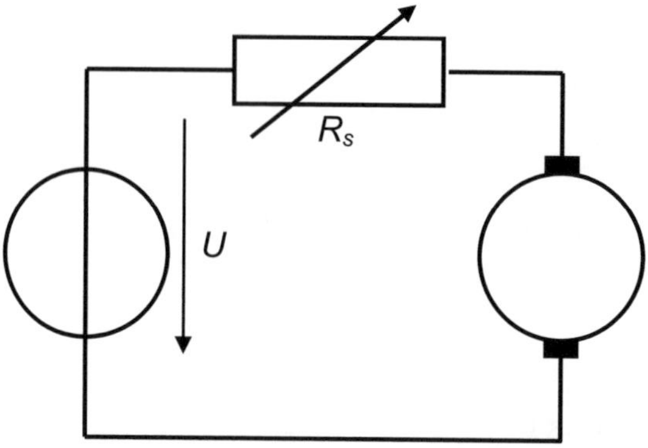

Figure 8.6. *Starting a DC motor with a rheostat in series*

As the torque and current characteristically decrease with the speed, resistance must generally be variable due to breaks (rheostat).

This technique is used for installations without speed that are controlled through the voltage, which are usually powerful motors (traction, rolling mills).

8.4.3. *Start-up inductance*

Reducing the start-up current of an induction motor can be performed by an impedance in series. *A priori* three possibilities exist (Figure 8.7):

– Resistance in series: losses caused in resistance are prohibitive and exclude such a solution.

– Capacitor in series: although this solution is interesting in terms of reactive power, it has the disadvantage of a large apparent capacitor power, so the cost is too high.

– Coil inductance in series: this is the most advantageous solution economically.

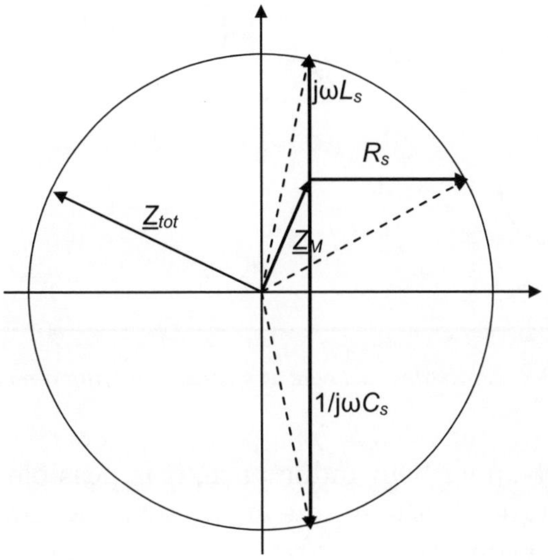

Figure 8.7. *Motor start-up with resistance, inductance or a capacitor in series*

Generally, the equivalent electrical scheme for a motor supplied by the mains, with inductance in series, can be reduced as follows, for one phase (representation in per unit values p.u.):

$\underline{z}_r$ = main internal impedance;

$\underline{z}_l$ = line impedance;

$\underline{z}_t$ = transformer short-circuit impedance;

$\underline{z}_L$ = inductance in series impedance;

$\underline{z}_M$ = motor impedance;

$\underline{u}_r$ = mains voltage;

$\underline{i}_{mL}$ = motor current with inductance;

$\underline{z}_{tot} = \underline{z}_r + \underline{z}_l + \underline{z}_t + \underline{z}_M + \underline{z}_L$;

$i_{mL} = u_r/z_{tot}$.

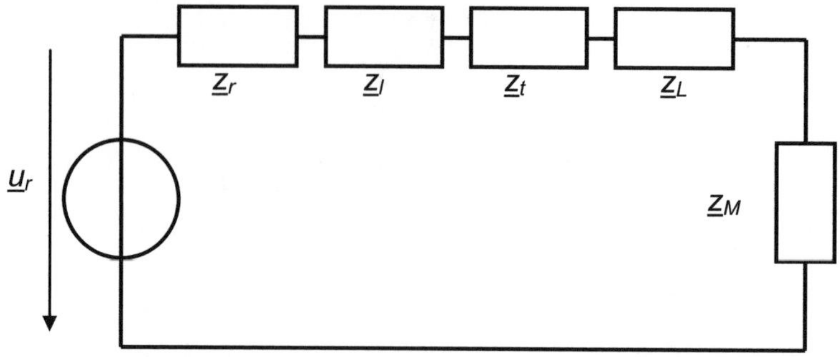

Figure 8.8. *Equivalent scheme of mains supplying a motor in p.u.*

For start-up without inductance, it is possible to refer the start-up torques to the torque at rated voltage M_{do}. It is thus possible to write:

– For the start-up without inductance:

$$M_d = M_{d0}\left(\frac{z_M}{\left|\underline{z}_M + \underline{z}_r + \underline{z}_l + \underline{z}_t\right|}\right)^2 \qquad [8.12]$$

– For the start-up with inductance:

$$M_{dL} = M_{d0}\left(\frac{z_M}{z_{tot}}\right)^2 \qquad [8.13]$$

8.4.4. *Example*

An induction motor has a 0.2 p.u. impedance at start-up and a corresponding torque at rated voltage M_{do} equal to 0.6 p.u. The objective is to reduce the current to a value of 2 p.u. For large-scale installations, most of the impedances can be considered as inductances. The other impedances have the following values:

$u_r = 1$

$x_r = 0.01$

$x_l = 0.01$

$x_t = 0.04$

$x_r + x_l + x_t + x_M = 0.26$

$i_M = 1/0.26 = 3.85$

For $i_{ML} = 2$, we have:

$x_{tot} = 0.5$

$x_L = 0.24$

The torques are, with and without inductance [8.12] and [8.13]:

$$M_d = 0.6\left(\frac{0.2}{0.26}\right)^2 = 0.355$$

$$M_{dL} = 0.6\left(\frac{0.2}{0.5}\right)^2 = 0.096$$

Under the same conditions and with the same constraints, using an autotransformer would have produced the following results:

$\ddot{u} = 1.483$

$M_{dat} = 0.211$

In this case, with the same main current, the torque is more than double its value with an inductance. The transformer maintains the power factor, while having an inductance in series decreases it.

Chapter 9

Electronic Peripherals

9.1. Power electronic

Any electrical drive requiring position or speed control or whose motor involves a non-electromechanical switching (synchronous or self-commutated motors) uses an electronic power device. It can go from a rectifier to a switching bridge. In addition to power electronics, regulation components (micro-processors, PSD, etc) and sensors are also used.

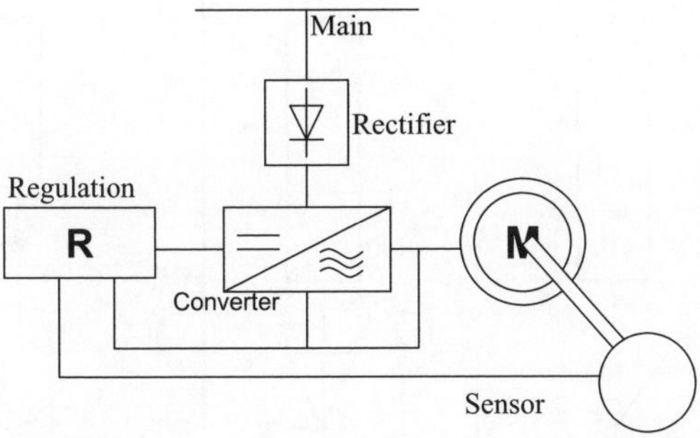

Figure 9.1. *Electrical and electronic components of an electric drive*

Electronic power devices used in the field of electric drives can be summarized in three basic configurations:

– simple switch;

– H bridge;

– 6 component bridge.

Basic electronic components are diodes, transistors (mainly IGBT), for very high power thyristors and TRIACs (triodes for alternating currents) for single-phase applications.

9.2. Simple switch

9.2.1. *Basic structure*

The *simple switch* is composed of a source, a coil and an electronic switch in series (Figure 9.2a).

In its simplest version, the switch is replaced by a diode (Figure 9.2b). It becomes a rectifier with one alternation if the source is an alternative source. In its active version, the switch consists of a transistor that operates in full (on) or blocked (off) mode (Figure 9.2c).

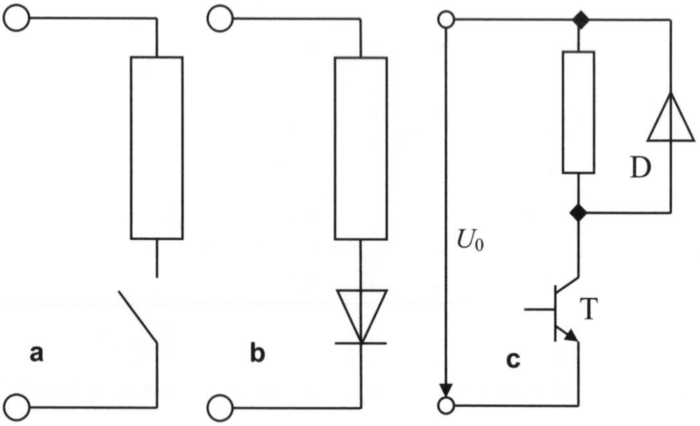

Figure 9.2. *a) Simple switch, b) diode in series, c) transistor switch with extinction diode*

9.2.2. *Active switch*

A transistor switch blocking operation virtually implies a current break and so an important over-voltage due to the winding inductance. This can be avoided by a diode in parallel on the winding (Figure 9.3a). During transistor blocking, inductive energy accumulated in the inductance is eliminated in the circuit resistance by circulating through the diode.

Figure 9.3b illustrates the case of a three-phase unipolar motor, for example the variable reluctance motor from Figure 4.38, powered by three active unipolar switches. The diodes lead to an extinction resistance to reduce the current extinction time constant in the phases.

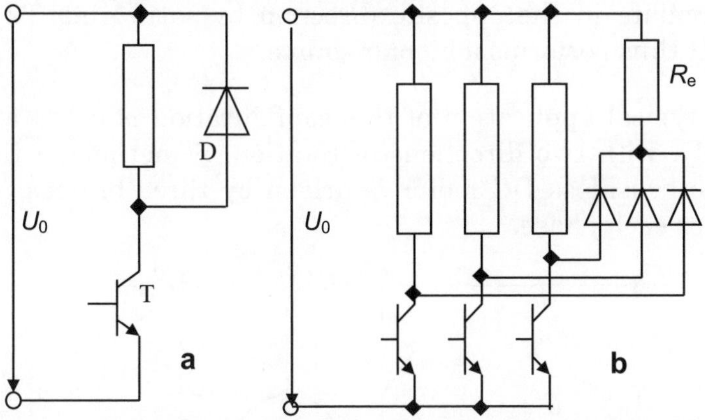

Figure 9.3. *a) Active switch with extinction diode, b) supply with three active switches and three extinction diodes leading to a common resistance R_e*

9.3. H bridge

9.3.1. *Basic structure*

The *H bridge* includes 4 switches whose topology is in the shape of an H (Figure 9.4a). To avoid any short-circuits, two switches located on the same vertical H branch cannot be on simultaneously. The simplest application of this structure is to replace the switches with four diodes, as seen in Figure 9.4b. This becomes a Graetz's bridge which, plugged into an alternative source, plays the role of a rectifier with double alternation.

9.3.2. *Active H bridge*

An *active H bridge* includes 4 transistors, protected against over-voltages by diodes mounted in anti-parallel. Figure 9.5 illustrates the principle.

From a DC source it is possible to let the current circulate in one direction by saturating transistors T1 and T2', and let it circulate in the opposite direction by saturating T2 and T1'. It thus performs a bipolar source.

A typical application of this configuration is a DC motor supply, with two directions of rotation. Sometimes a three-phase brushless DC motor is driven by three bridges in H, one for each phase.

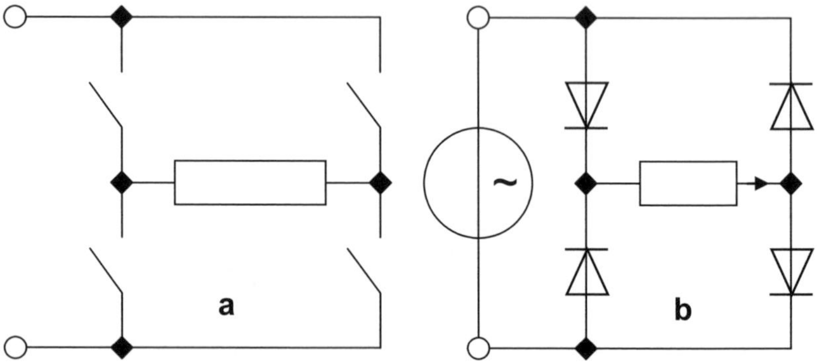

Figure 9.4. *a) H bridge with 4 switches, b) H rectifier or Graetz's bridge*

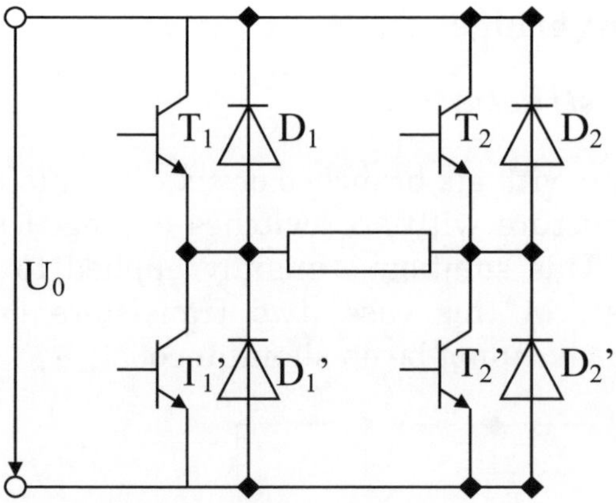

Figure 9.5. *Active H bridge with 4 transistors and 4 protection diodes*

9.3.3. *Half-H bridge*

The *half-H bridge* is characterized by two transistors and two diodes (Figure 9.6). The current circulation in load has only one direction, the same as the active switch. However, the difference with a simple switch is the possibility of recovering energy when blocking a transistor.

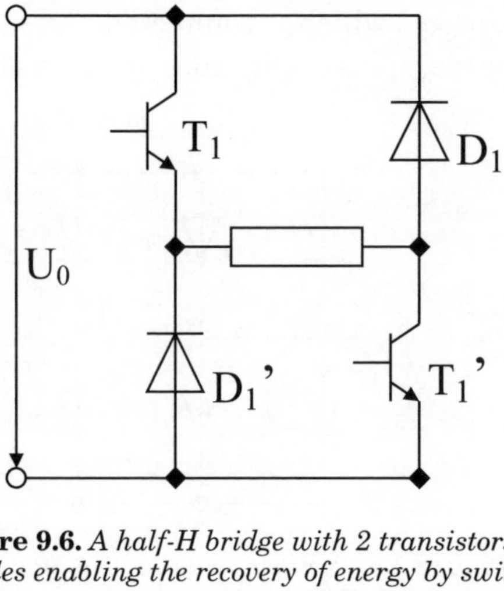

Figure 9.6. *A half-H bridge with 2 transistors and 2 diodes enabling the recovery of energy by switch off*

9.4. Element bridge

9.4.1. *Basic structure*

The bridge with six branches or six elements is defined in its basic structure with six switches arranged according to Figure 9.7. This solution is mainly applied to three-phase motors. Also, in this case, two transistors on the same vertical branch cannot be on simultaneously.

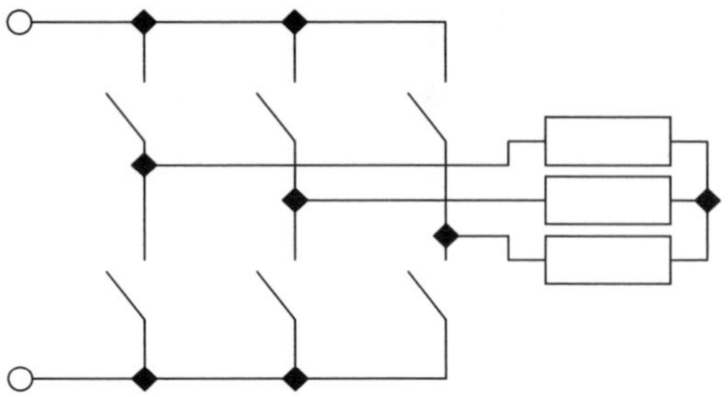

Figure 9.7. *Bridge with 6 branches supporting 6 switches*

The simplest application of this scheme is to replace the switches with six diodes, according to Figure 9.8. It then has a *rectifier bridge* which, connected to a three-phase alternative source, plays the role of a rectifier with six alternations.

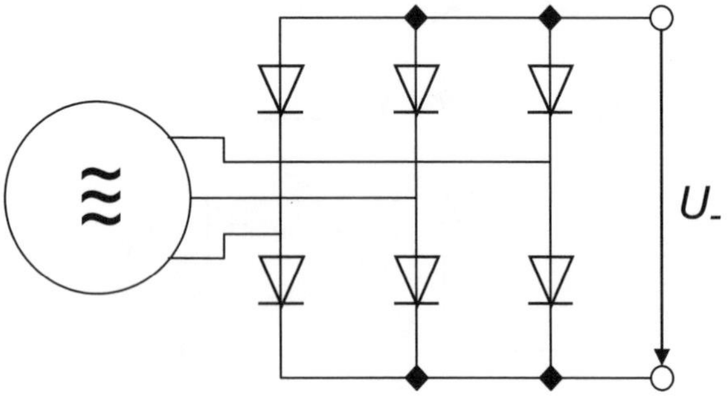

Figure 9.8. *Rectifier bridge with 6 diodes*

9.4.2. *Six transistor active bridge*

An *active bridge* is obtained by replacing switches with transistors (or other active elements as thyristors), themselves protected against over-voltage by diodes mounted in anti-parallel (Figure 9.9).

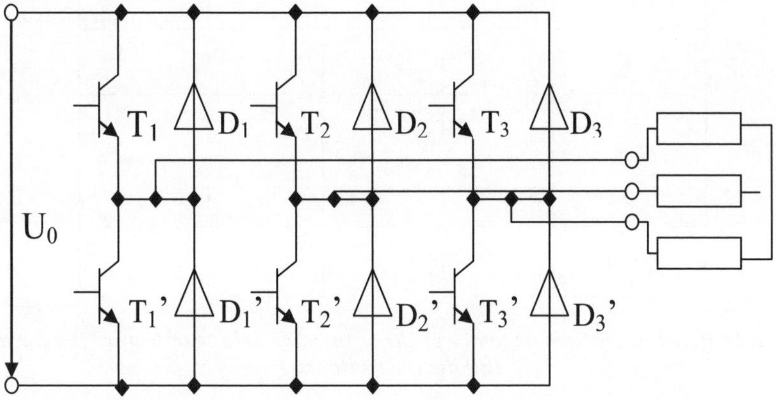

Figure 9.9. *Active bridge with 6 transistors and 6 diodes*

Various possibilities of transistor switching are possible, from 6 logical statements per period to a PWM intended to generate a three-phase sinusoidal system. Later we will describe two frequently used switching modes: 120 degree and 180 degree commutations.

9.4.3. *120 degree commutation*

120 degree commutation implies that, at each switching sequence of two transistors, one on an upper branch and the other on a lower branch are saturated simultaneously. This sequence is summarized in Table 9.1 and represented in Figure 9.11.

Sequence	Upper transistor	Lower transistor
1	T_1	T_2'
2	T_1	T_3'
3	T_2	T_3'
4	T_2	T_1'
5	T_3	T_1'
6	T_2	T_2'
7 = 1	T_1	T_2'

Table 9.1. *Saturation sequence of the transistor bridge of Figure 9.9 for a 120 degree switching*

Figure 9.10 describes the succession of logical states of the 6 transistor bridge of Figure 9.9 for a period, with the following convention:

0 = blocked transistor;

1 = saturated transistor.

The resulting phase voltage is also represented at the bottom of the figure. It shows a form of periodic square wave with a third period width, hence the designation of a 120 degree commutation. Between a positive wave and a negative wave, a 60 degree dead time appears.

For a supply state, a positive current appears in one phase and the same negative current appears in a second phase. The third phase current is zero.

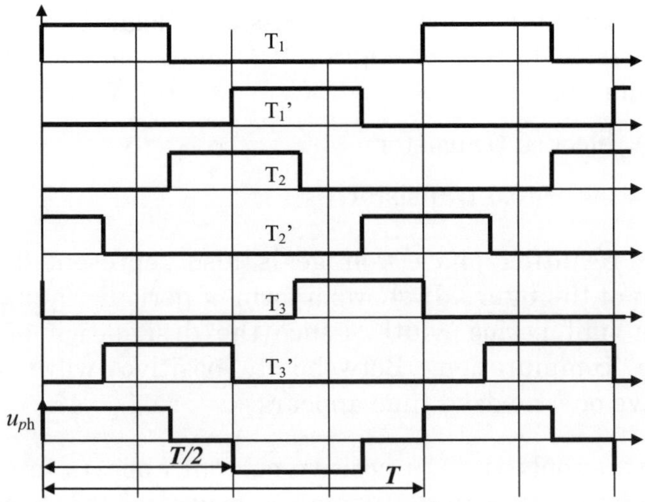

Figure 9.10. *Logical state of the transistors and voltage phase on a period for a 120 degree commutation*

9.4.4. *180 degree commutation*

180 degree commutation implies that, at each switching sequence of three transistors, one on an upper branch and the other two on a lower branch or vice versa are saturated simultaneously. This sequence is summarized in Table 9.2 and represented in Figure 9.11.

Sequence	Transistors 1	Transistors 2	Transistors 3
1	T_1	T_2'	T_3
2	T_1	T_2'	T_3'
3	T_1	T_2	T_3'
4	T_1'	T_2	T_3'
5	T_1'	T_2	T_3
6	T_1'	T_2'	T_3
7 = 1	T_1	T_2'	T_3

Table 9.2. *Saturation sequence of the transistor bridge of Figure 9.9 for a 180 degree commutation*

Figure 9.11 describes the succession of logical states of the 6 transistor bridge of Figure 9.9 for a period, with the following convention:

– 0 = blocked transistor;

– 1 = saturated transistor.

The resulting phase voltage is also represented at the bottom of the figure. It shows a form of periodic square wave with a half period width, hence the designation of a 180 degree commutation. Between a positive wave and a negative one, no dead time appears.

For a supply state, a positive current appears in a phase and the half negative current appears in the two other second phases or vice versa.

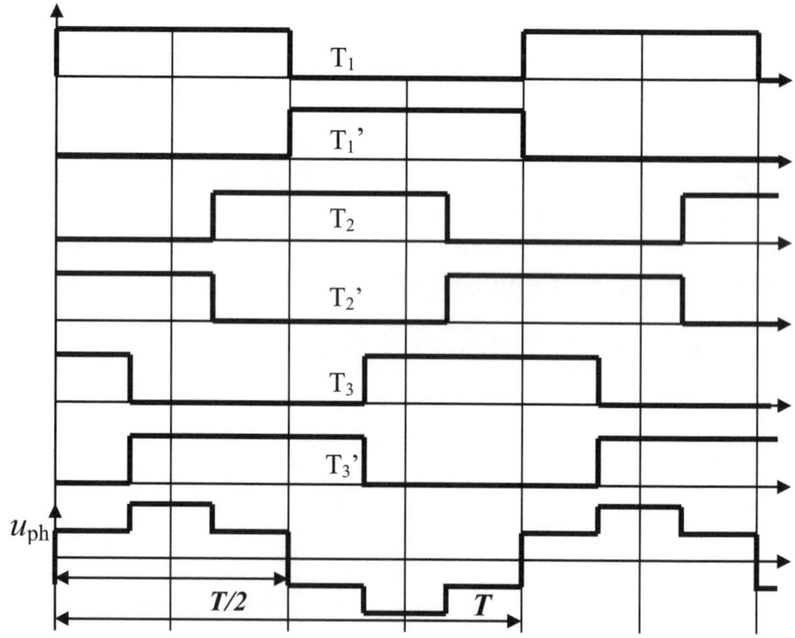

Figure 9.11. *Logical state of the transistors and voltage phase on a period for a 180 degree commutation*

Chapter 10

Sensors

10.1 Functions and types

10.1.1. *Functions*

Sensors to control an electric drive mainly fulfill the following functions:

– Measurement of quantities required by the motor management such as current in order to control torque and rotor position for the commutation of a converter to generate the phase supply.

– Measurement of quantities required by the driven body management, such as position and speed in order to meet specifications related to the application.

– Measurement for the electric drive protection, in particular against over-heating, over-speed or locking.

10.1.2. *Position and speed*

Generally, the position and speed functions are performed by the same sensor. It is characterized by the following properties:

– the *resolution*, defined by a number of increments per revolution (rotating system) or by the elementary step value (linear system);

– *accuracy* and speed range measurement is frequently defined by the sensor and the signal processing system, in particular for digital sensors;

– the possibility for a position sensor to provide an absolute or relative value;

– the possibility for a position sensor to provide information at standstill.

10.1.3. *Sensor types*

For position and speed measurement systems, there are mainly 5 technologies:

– optical sensors;

– inductive sensors;

– magnetic sensors such as *Hall sensors* or *magneto-resistances*;

– capacitive sensors;

– indirect sensors or observers.

Three techniques are primarily used for current measuring:

– measuring the voltage at resistance terminals, often in the form of an integrated sensor;

– current transformer for AC high current level;

– indirect current measurement by a magnetic field and a Hall sensor.

For thermal measurements, typically two techniques are used:

– *thermo-resistances* for a critical threshold detection, requiring an alarm or shutdown;

– *thermocouples* for precise continuous measurement.

10.2 Optical position sensors

10.2.1. *Principle*

The *optical position and speed sensor* solution is used frequently. The principle is based on the degree of brightness by the photo-electric effect. Two key techniques are used:

– Measurement by transmission between a transmitter (LED, laser) and a receiver (photodiode, phototransistor), using a disk with alternately transparent and opaque areas. The disc mounted on the shaft consists of a punched metal for a technology with low or medium resolution (Figure 10.1) and by a glass engraved disk for high resolutions.

– Measurement by reflection (Figure 10.2). The light path modulation is obtained by a structure (disk or ruler) which alternates absorbent and reflective areas.

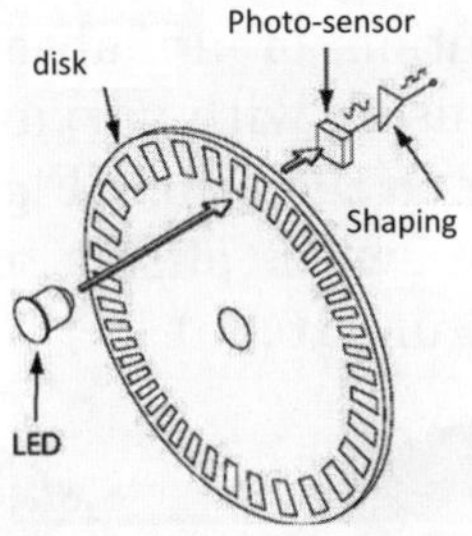

Figure 10.1. *Incremental optical position and speed sensor by transparency*

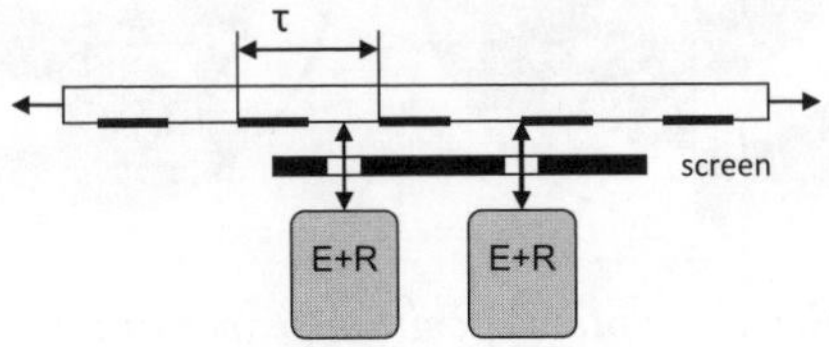

E+R =transceiver

Figure 10.2. *Incremental optical position and speed sensor by reflection*

The sensor system is usually double, with an offset of $n\tau + \tau/4$ between the two parts (Figure 10.2), with τ the period of alternation between opaque and transparent elements. This allows us to increase the resolution by a factor of 4 and the detection of rotation direction.

10.2.2. *Performances*

Cheap optical sensors reach a resolution of 1,000 to 2,000 increments per revolution, using technology by transparency.

High resolutions, up to 100,000 increments per revolution, are reached by using engraved glass discs. These systems typically require a second runway at lower resolution to measure speed for high values.

For measurements associated with movements of translation, rulers offering the same optical detection principle exist.

For the determination of an absolute position, several concentric tracks are used, with varying resolutions in a 1 to 2 ratio from one track to the other (Figure 10.3a) or with the same resolution, but with offsets corresponding to the desired resolution (Figure 10.3b).

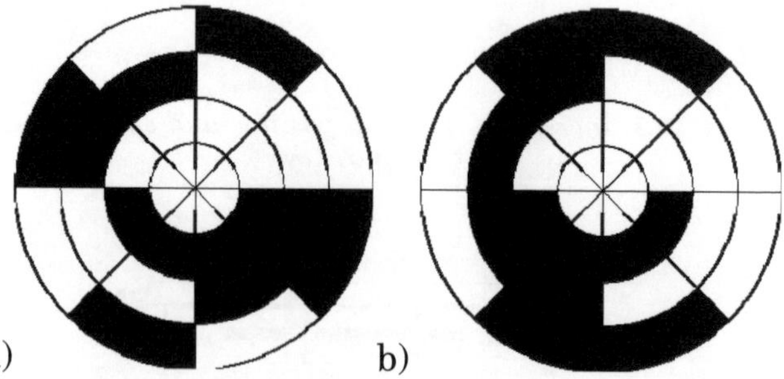

a) b)

Figure 10.3. *a) Absolute optical position sensor and b) variable resolution for each track; with the same resolution for each track but with offset corresponding to the final resolution*

10.3. Hall sensors

10.3.1. *Principle*

A *Hall sensor* is an electronic component box shape, supplied in current in one direction. When submitted to a magnetic induction in a perpendicular direction to the largest surface (Figure 10.4), a voltage difference proportional to that field appears in the third direction.

Such a probe is therefore able to detect a local magnetic induction field in direction and intensity.

10.3.2. *Applications*

A frequent use, associated with the rotor of a brushless DC motor, is to detect the position of each pole transition. The use of a number of probes equal to the number of phases and shifted between them by a third of a period makes it possible to directly generate phase commutation for power supplies to 120 or 180 degrees (Figures 9.11 and 9.13).

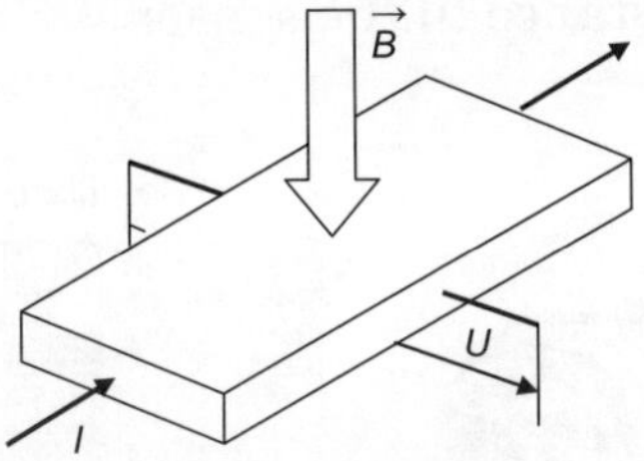

Figure 10.4. *Hall sensor*

A position sensor can also be achieved by means of a cylinder-shaped or multipolar disk auxiliary magnet acting with Hall probes. This method is cheap but only suitable for low resolutions.

Hall sensors also allow the indirect measure of currents (see section 10.8.4).

10.4. Inductive position sensors

10.4.1. *Principle*

An *inductive sensor* works by interaction of an AC magnetic field source with a ferromagnetic or conductive element. The corresponding frequencies generally range from 1 to 100 kHz. The apparent impedance of the source sensor changes due to feed-back in front of an object to be detected, ferromagnetic or conductive (eddy current appearance) nature, and thus enables detection based on a reference threshold.

10.4.2. *Applications: simple sensor with variable self-inductance*

Such an inductive sensor is described in Figure 10.5. It consists of a ferrite-type ferromagnetic circuit and an associated coil, powered by an AC voltage or current source. The proximity of a metal object changes the apparent impedance and can thus be detected with a level which varies with the distance to the sensor.

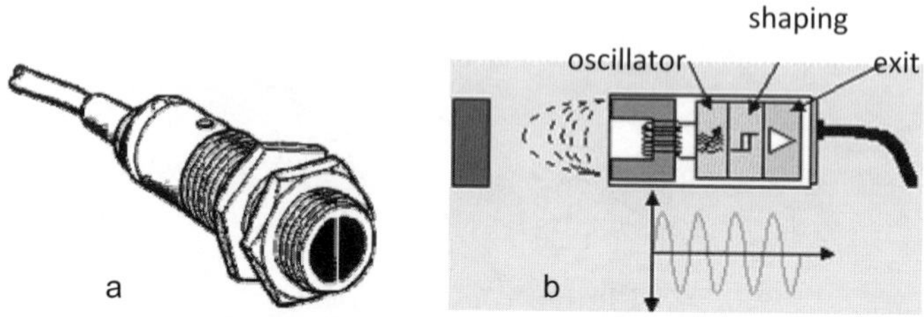

Figure 10.5. *Inductive proximity sensor: a) probe; b) structure*

A possible application for electric drives is to associate them with a toothed wheel mounted on the motor shaft or driven body (Figure 10.6). Tooth-notch or notch-tooth transition creates a transition signal which defines position information.

Figure 10.6. *Inductive proximity sensor associated with a toothed wheel*

10.4.3. *Linear variable differential transformer*

The *linear variable differential transformer or LVDT sensor* is based on an inductive coupling between an emitter coil (a) and two measurement coils (b1, b2) mounted in opposition, associated with a mobile ferromagnetic core (c, usually ferrite). Figure 10.7 describes the principle. The measuring voltage U_{meas} is the voltage difference of the respective back-EMF of coils b1 and b2. In particular, for a centered position of the core, the U_{meas} voltage is zero.

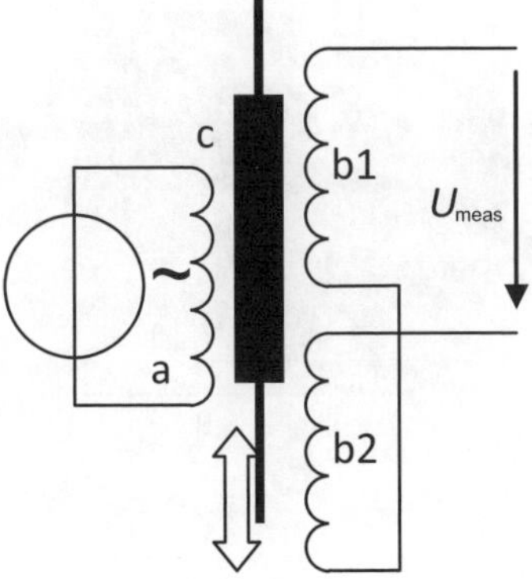

Figure 10.7. *LVDT sensor – principle*

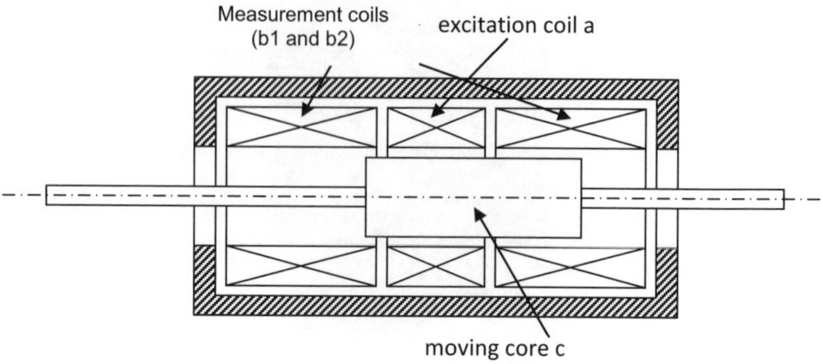

Figure 10.8. *LVDT sensor – structure*

Figures 10.8 and 10.9 describe the structure of such a sensor. The moving core is ferrite, the position of which affects the coupling between a coil and two b coils is represented. Figure 10.10 shows the evolution of the signal amplitude U_{meas} depending on the core position, along with its phase. Figure 10.11 illustrates this type of sensor picture with the 3 coils for the generator circuit A and measurement circuit B.

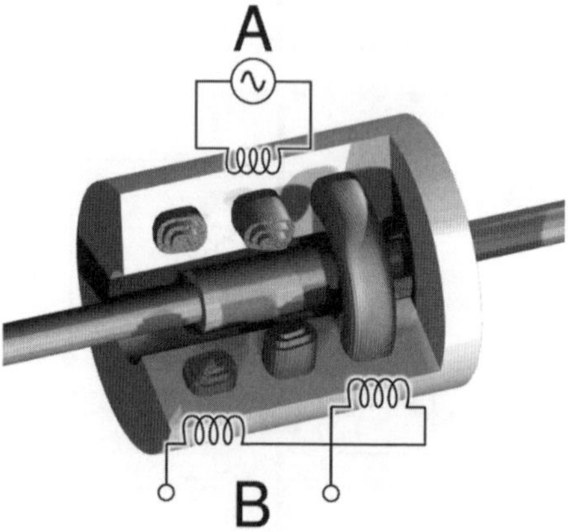

Figure 10.9. *Linear LVDT sensor – structure*

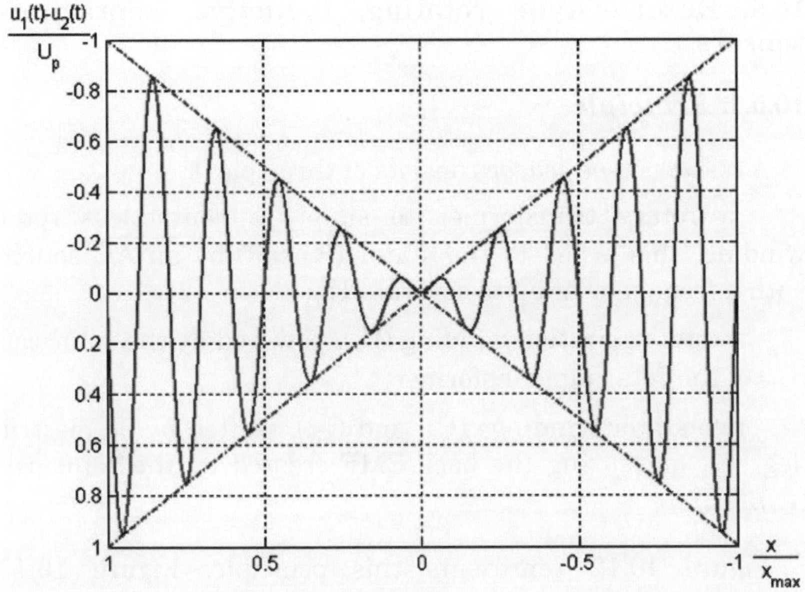

Figure 10.10. *LVDT sensor – signal measurement, amplitude and phase*

Figure 10.11. *LVDT sensor – system, generator circuit and shaping*

10.5. Resolver-type rotating, inductive, contactless sensors

10.5.1. *Principle*

A *resolver type sensor* consists of three parts:

– a rotary transformer to supply a contactless rotor winding, this is fed to the stator (E_s coil) by an AC source with a frequency of up to several kHz;

– a rotor excitation winding (E_r), powered by the rotor coil (C_r) of the rotating transformer;

– two stator windings (C_{s1} and C_{s2}), shifted by 90 electric degrees, measuring the back-EMF created by the rotor coil (C_r).

Figure 10.12 represents this principle. Figure 10.13 describes the structure and the arrangement of the components.

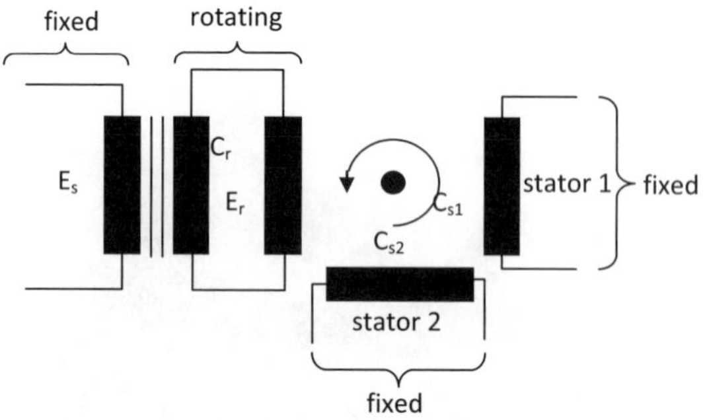

Figure 10.12. *Resolver sensor principle: rotating transformer (E_s,C_r), rotor excitation coil (E_r) and measuring signal capture coils (C_{s1},C_{s2})*

Figure 10.14 shows such a sensor, with separated and assembled stator and rotor. Figure 10.15 shows the typical speed signals. Through signal demodulation, it is possible to infer the absolute position of the axis, at any speed, including when it is stationary.

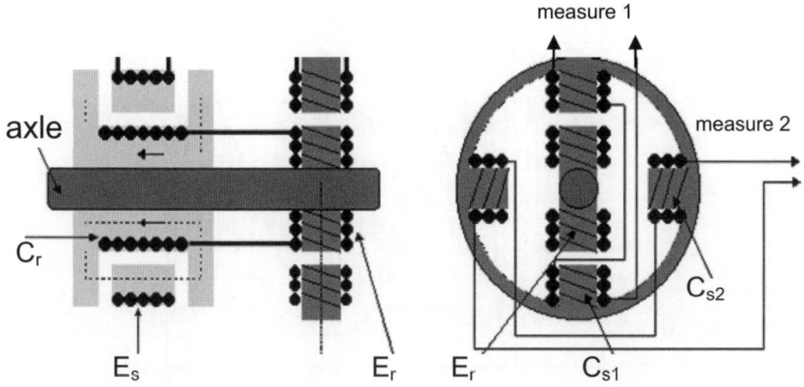

Figure 10.13. *Resolver sensor structure: rotating transformer (E_s,C_r), rotor excitation coil (E_r) and measuring signal capture coils (C_{s1},C_{s2})*

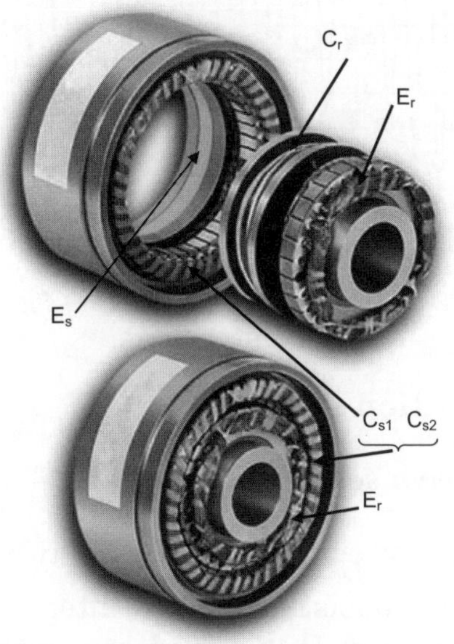

Figure 10.14. *Resolver sensor – stator, rotor and assembling*

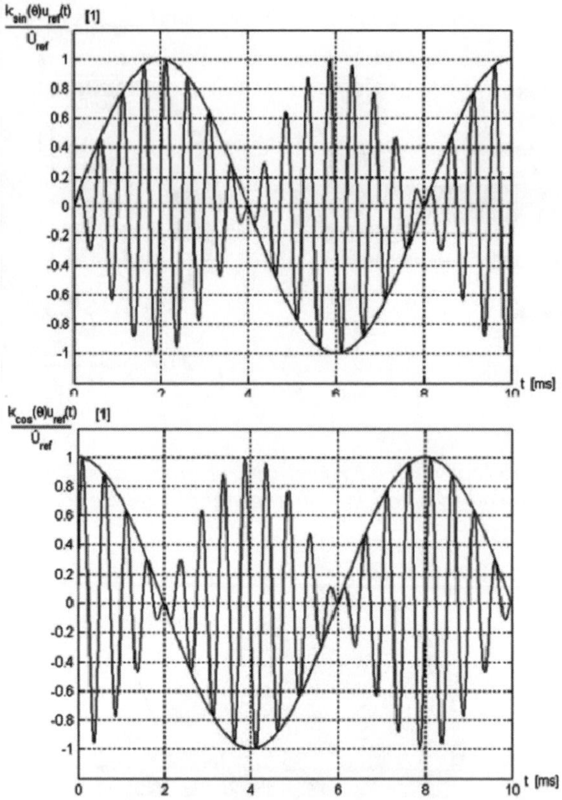

Figure 10.15. *Resolver sensor – coil measuring signals shifted by* $\pi / 2$

10.6. Other position sensors

10.6.1. *Inductosyn sensors*

An *Inductosyn*-type sensor consists of two parts:

– a fixed part consisting of two integrated circuits in a comb shape, typically PCB, fixed and powered by an AC source of several kHz;

– a mobile part, consisting of a PCB winding with the same pitch as the fixed part, picking up the mutual back-EMF generated.

Figure 10.16 illustrates the principle for a linear structure. One such sensor can be incremental or absolute; it may be linear (Figure 10.17) or rotary (Figure 10.18). Its particularity is that it does not contain ferromagnetic elements, thus eliminating any non-linearity. However, such a sensor is sensitive to magnetic interferences.

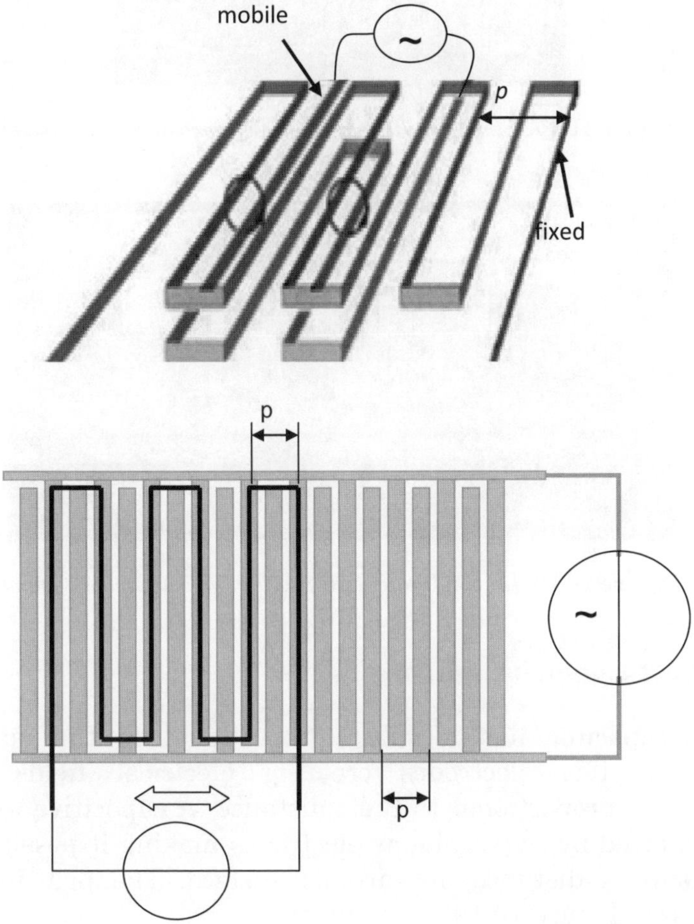

Figure 10.16. *Inductosyn position measurement system – principle – perspective and plan views*

Figure 10.17. *Linear Inductosyn position measurement system*

Figure 10.18. *Linear Inductosyn position measurement system*

10.6.2. *Capacitive sensors*

A capacitor has a value proportional to the surface between two electrodes creating electrical fields and inversely proportional to their distance. A capacitive sensor is achieved by two coplanar electrodes making it possible to measure a distance or variable surface. The principle is shown in Figure 10.19.

To measure a position, electrodes with varying surfaces, based on their relative positions, are used. Figure 10.20 shows an example of a rotating sensor. The left part is fixed and the right, linked to an axis, is rotating. Tracks c and d are supplied in AC (in the order of MHz). They polarize the rotating tracks e and f by an electric field. Tracks a and b can pick up a variable voltage difference depending on the position. Figure 10.21 describes the bridge circuit used to obtain the measurement. Such sensors can be achieved by the principle of integrated circuits. They require a perfect parallelism of fixed and mobile parts.

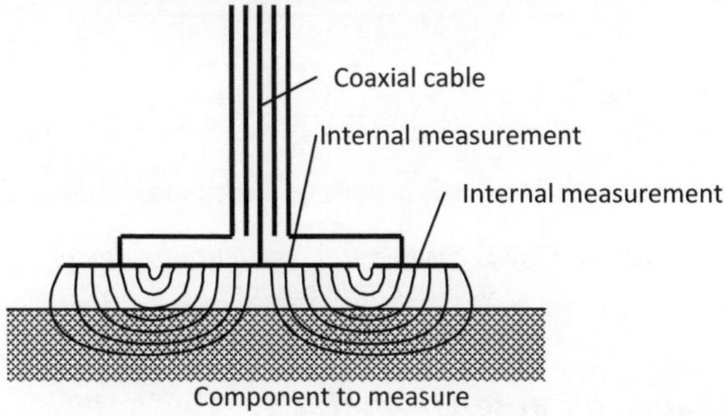

Figure 10.19. *Capacitive measuring principle*

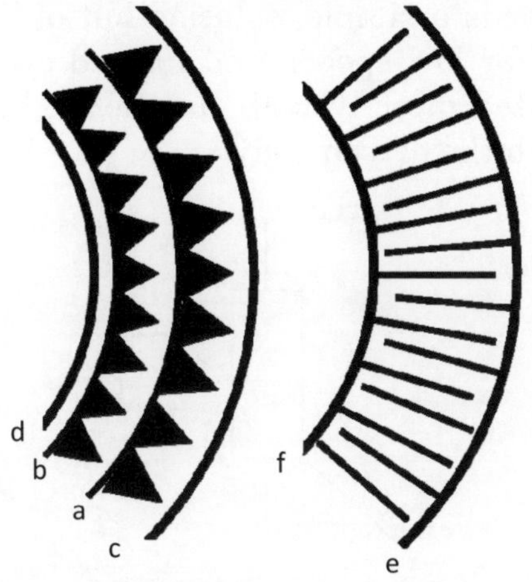

Figure 10.20. *Fixed (left) and mobile (right) electrodes of a rotating capacitive sensor*

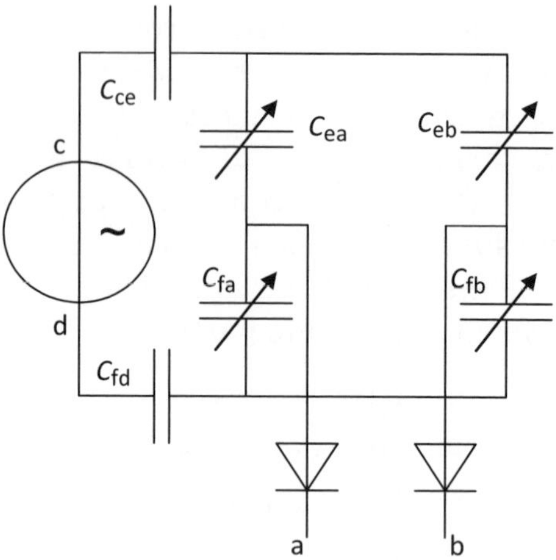

Figure 10.21. *Supply and measure bridge of a capacitive sensor*

10.6.3. *Sensors by potentiometer*

Using a potentiometer resistance for which the cursor is associated with the mobile part (Figure 10.22), a linear or rotating sensor is a simple, solution but of limited use. It is only suitable for low speeds and limited cycles (wear). It is rarely associated directly with the motor, but is sometimes associated with the driven body.

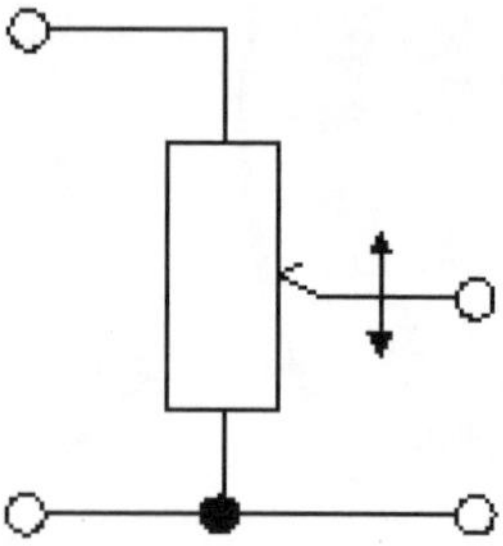

Figure 10.22. *Position sensor by a potentiometer*

10.7. The motor as a position sensor

10.7.1. *Principle*

All motors have a number of recurring defects associated with the stator, the rotor or interaction of both:

– tooth-slot alternation;

– saturated zones, varying with the rotor position;

– field harmonics;

– back-EMF periodicity;

– anisotropic character of materials (PM), etc.

The idea of an *indirect position sensor* using the motor itself is to advantageously use these intrinsic defects. If it is not possible to obtain a high resolution by such a process, however, a sufficient resolution for a speed or commutation control of a brushless DC motor is possible. The main advantage is the reduction of mechanical components and the absence of sensor positioning relative to the rotor (which is then implicit), which is always difficult. Some possibilities are described below.

10.7.2. *Back-EMF*

Any permanent magnet or variable reluctance motor shows back-EMF that can be directly measured at no load in rotation. For the case of a permanent magnet rotor, the back-EMF phase voltage in rotation is:

$$u_{iph} = k_u \Omega . \sin p\Omega t = k_u \Omega \sin \alpha(t)$$

This voltage contains both position and speed information. In particular, zero crossing of this voltage defines specific rotor positions with the corresponding phase. Specifically, over a period (a polar pitch), the three phases of a brushless DC motor have 6 zero crossings spaced by a sixth period. A

120 degree supply also requires 6 commutations per period. During the interval for which a phase is not supplied (1/6 of period, Figure 9.11), it is possible to measure the back-EMF, especially its zero crossing which coincides with the middle of this interval. By introducing a switching delay of 1/12 of a period, it is possible to use this signal to switch the corresponding phase. Figure 10.23 shows this possibility. The limitations of this technique are the impossibility of detecting a signal without movement and a resolution limited to the polarity as follows:

$$N_p = 6p$$

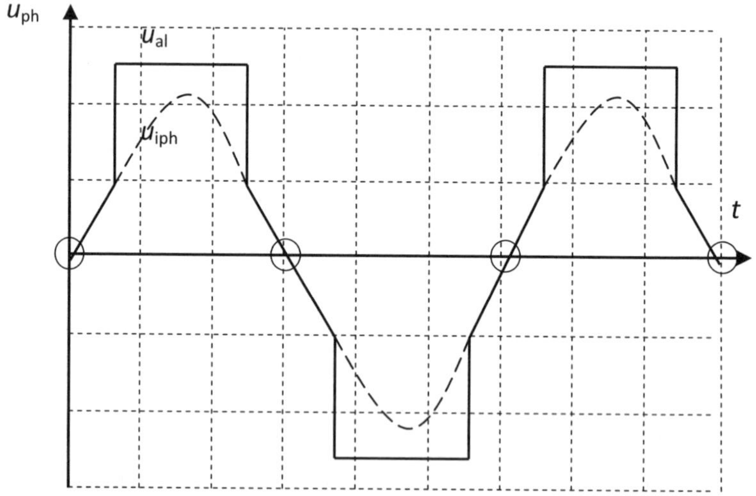

Figure 10.23. *Back-EMFe u_{iph} and supply phase voltage u_{al} for a 120 degree brushless DC motor – zero crossing used for commutation*

10.7.3. *Saturation level measurement*

Any permanent magnet motor is subject to a magnetic flux created by the magnet, even to standstill. The result is a *saturation* state of the teeth that depends on the position.

An indirect measure of this report provides position information. This determination can be achieved by sending a short pulse voltage on one phase, first positive, then negative. The flux resulting from the phase current is added to or subtracted from the magnet flux. If it is added, saturation increases and inductance decreases and the time constant linked to the phase decreases. If the flux is subtracted, the inductance and the time constant increase. If there is a successively positive voltage pulse of calibrated duration, then a negative one sent on a phase coil, there are two possible cases:

– the positive current pulse is higher than the negative current pulse; positive flux has the same sign as the magnet flux, the negative flux is in the opposite direction: the phase is in front of a North pole;

– the negative current pulse is higher than the positive current pulse; negative flux has the same sign as the magnet flux, the positive flux is in the opposite direction: the phase is in front of a South pole.

By repeating this procedure for each of the 3 phases, it is possible to determine the position of the rotor from the stator with an accuracy of $1/6$ of a period, which is enough to energize the phases with a maximum torque in the desired direction of rotation.

If the measuring conditions are those of Figure 10.24, positive and negative pulses appear in three phases, for a given position of the rotor. These impulses have the same apparent sign, as those measured from a continuous source level. There is a successively negative pulse. The amplitude of phase differences are successively negative for phase 1, positive for phase 2 and negative for phase 3. So, phase 1 is in front of a South pole, phase 2 opposite to a North pole and phase 3 to a South pole. It is thus possible to choose the optimal supply sequence. Reference [CAR 93] describes the process in detail.

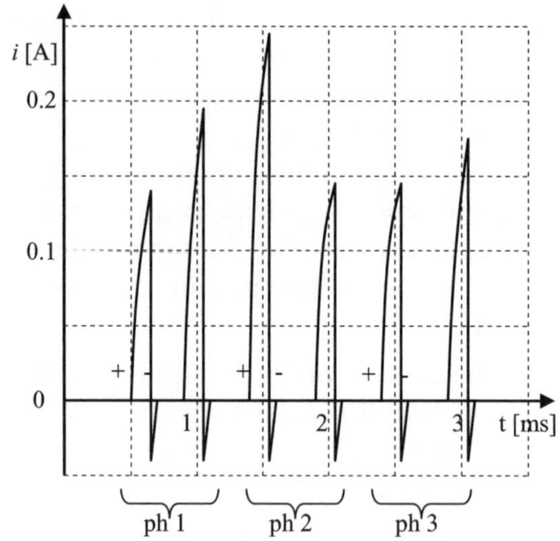

Figure 10.24. *Current pulses of a three-phase motor with PM; indirect measurement of the flux direction created by the magnet in the different phases*

10.7.4. *Detection of the third harmonic*

For most motors, flux rotor harmonics create a back-EMF *third harmonic* in the different phases. An ungrounded (without neutral to the ground) star coupling allows us to eliminate the multiples of three current harmonics. The corresponding harmonic flux and back-EMF nevertheless remain (Figure 10.25). It is possible to measure these voltage harmonics using the process described in Figure 10.26. Applying voltage supply to a three equal resistance bridge mounted in parallel with the motor makes it possible to create a second neutral point N_2. By measuring the voltage difference between N_2, and the motor neutral N_1, it is possible to eliminate the fundamental voltage and to keep only the harmonics that are multiples of three, generated by the motor.

The third harmonic presents 6 zero crossings over the time period, enabling us to ensure a brushless DC motor commutation. Reference [OSS 88] describes in detail the process, in particular the synchronization at start-up.

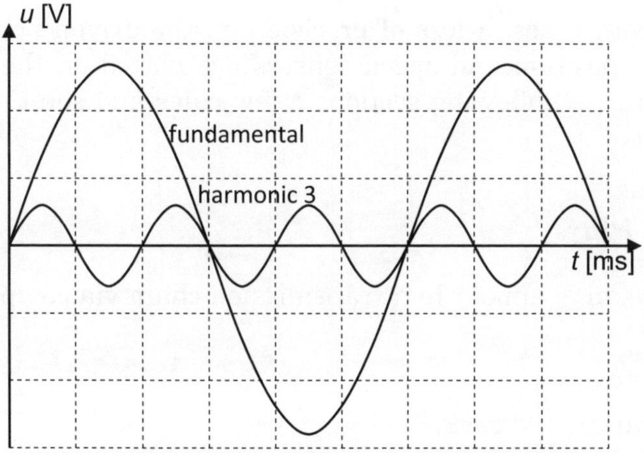

Figure 10.25. *Fundamental and third harmonic of a PM motor back-EMF*

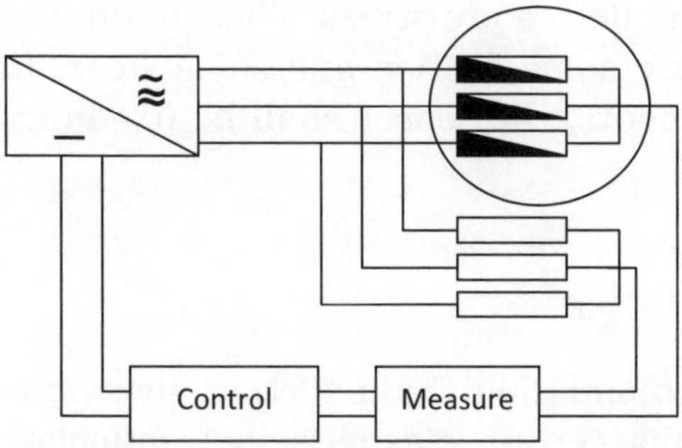

Figure 10.26. *Back-EMF third harmonic detection process*

10.8 Sensor position

10.8.1. *Problems*

Any position sensor can theoretically be placed at any point in the transmission chain. However two problems limit this possibility:

– *plays* that may appear in the transmission chain;

– the system *elasticity*, in traction or torsion.

In both cases, a loss of precision on the driven body may occur if position and speed sensors are placed on the motor shaft. In the following sections a few rules and solutions are proposed.

10.8.2. *Plays*

Plays may appear in a transmission chain via components such as:

– gears;

– gearing and rack;

– screws; etc.

There are systems eliminating the plays consisting of a corresponding device pre-stress. Their main drawback is that they introduce some friction and some wear. In the absence of these elements, the sensor shall be fixed directly onto the driven body.

10.8.3. *Elasticity*

A long transmission chain such as screws, belts or cables may present a certain *elasticity* in traction or torsion that creates a difference in actual position between motor and driven body measurement.

As for plays, the solution is to place the position and speed sensors directly onto the driven body. There remains the problem of the chain dynamic behavior that can lead to significant oscillations and even resonance phenomena. It is the control device which should be designed taking into account these aspects.

10.9. Current sensors

10.9.1. *Principle*

Direct *current measurement* is not possible. Various processes enable us to measure a current indirectly:

– measuring voltage at the terminals of resistance in series with the current;

– using a current transformer in AC;

– using a Hall sensor.

10.9.2. *Resistance in series*

A measure by a resistor in series necessitates a low value, to limit losses, while maintaining a measurable level value. In order to avoid a resistance variation, it must be chosen so as not to become significantly hot. It will nonetheless remain affected by external temperature variations. Resistance measurement is frequently not grounded. It must therefore rely upon a differential measurement voltage from operational amplifiers.

This technique can be used in DC as well as in AC. Some integrated circuits for power bridges include the resistance with the voltage circuit measurement.

10.9.3. *Current transformer*

Important alternative currents may require the use of a *current transformer*. It is characterized by a low iron saturation level and functions practically with a short-circuited secondary winding. It has the same characteristics as an electrical decoupling with a secondary winding as opposed to the primary winding, thus allowing the circuit to be grounded.

10.9.4. *Current measurement per Hall probe*

The principle of the Hall probe was described in section 10.3.

Accurate current measurement eliminating any non-linearity related to the measurement system is based on an open ferromagnetic structure surrounding the conductor associated with the current to be measured (Figure 10.27).

The Hall sensor detects the induction resulting from this current. With measurement circuitry, a current is generated and applied to a multi-turn winding surrounding the magnetic circuit to cancel the magnetic flux density in it at any time. Thus, the generated current is a picture of the current to be measured (Figure 10.28). The scale factor is defined by the number of coil turns associated with the Hall probe. It is possible to measure both AC and DC currents.

[FAV 04] describes the process in detail.

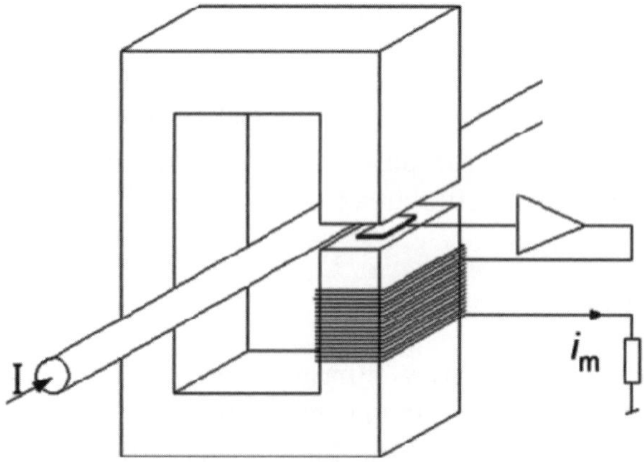

Figure 10.27. *Current measuring technique with a Hall probe – principle*

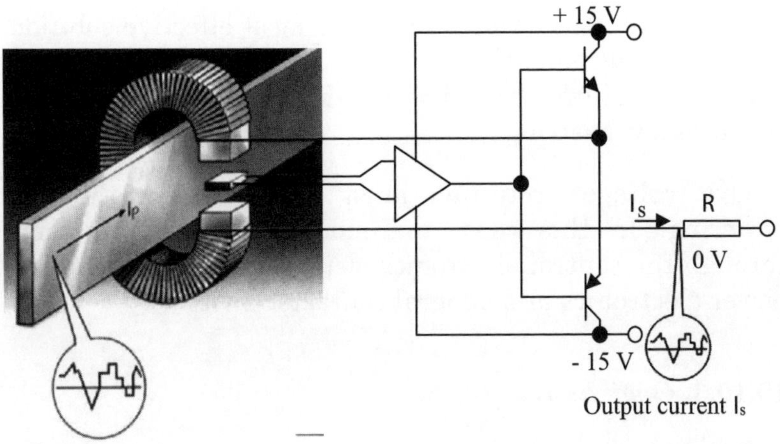

Figure 10.28. *Current measuring technique with Hall probe – measuring circuit*

10.10. Protection sensors

10.10.1. *Aim*

Any electric drive, and in particular any motor, must be protected against operational anomalies.

The following categories can be noted:

– over-current protection;

– over-voltage protection;

– over-heating protection;

– protection against water;

– protection against explosions.

10.10.2. *Over-current and over-voltage*

Over-current protection can be provided by a fuse, but this solution has the disadvantages of a period of installation shutdown and often an overly slow reaction for effective protection of power electronics. The most effective solution is to use measurements associated with electronic command which, for failure, acts directly on electronic power at the required level setting.

Over-voltage protection must also preserve the motor powering. In this case, the more effective protection is through the control electronics, acting where appropriate, on power electronics or a general contactor switch.

10.10.3. *Over-heating*

Over-heating protection must be provided for power electronics and motors.

Generally, for power electronics, a temperature sensor is used such as thermocouples placed at key places. While it is easy to act in time using current reduction or stopping in the case of insufficient cooling down, it is more difficult to intervene in time in the event of a short-circuit. Only command and power electronics are able to respond quickly.

For motors, they are generally built with temperature sensors included in the stator slots. These are usually thermo-resistances, rarely thermocouples. Motor thermal time constants range from several seconds (small motors with the winding in the gap) to tens of minutes (very large motors), enabling an adapted response.

10.10.4. *Other types of protection*

Other types of protection are sometimes required and are ensured by the motor design and construction. Among them are the degree of *water protection*, which can go up to the guaranteed operation of a fire hose (Figure 10.29) and submerged pump motors (Figure 10.30).

For electric drives operating in an atmosphere that could contain flammable gases, it is necessary to achieve an explosion-resistant *anti-deflagration* motor.

Figure 10.29. *Motor protected against fire hose*

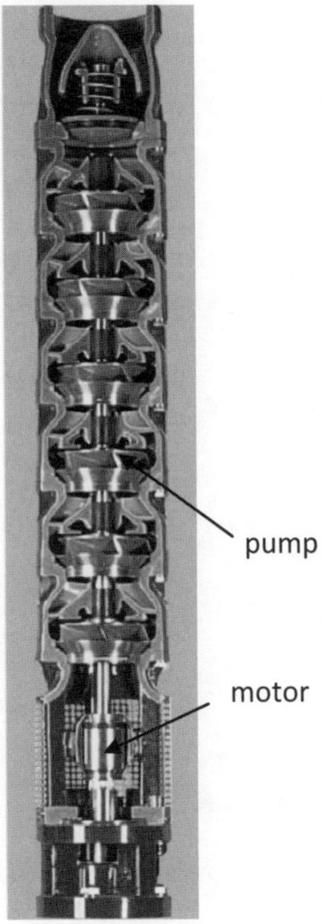

Figure 10.30. *Pump and submerged induction motor*

Chapter 11

Direct Drives

11.1. Performance limits

11.1.1. *Methodology*

Generally, a *direct drive* motor, either rotary or linear, can first be characterized by its surface force [JUF 99]. Even if this force is not frequently used, it is the most practical for characterizing such motors.

If τ is the surface tangential force, the linear force is then simply given by the expression:

$$F_t = \tau \cdot S_a \qquad [11.1]$$

S_a = active surface

For a rotary motor, the torque is given by:

$$M = \tau \cdot S_a \, d/2 \qquad [11.2]$$

d = boring diameter

For a given motor type, the performances will depend initially on the following values:

– the pole pitch for a linear motor;

– the boring diameter for a rotary motor.

The mode of cooling – natural cooling, forced ventilated cooling, water-cooling, etc. – will strongly influence the performances. The orders of magnitude which are fixed thereafter are given for a motor in air cooling mode.

Two cases will mainly be analyzed:

– the PM motor of synchronous self-commutated type, with control of the commutation angle;

– the induction motor, with short-circuited rotor cage.

11.1.2. *Specific surface force*

For a rotating or linear motor, the performances can be characterized by the *sheer stress force on the surface*. This force can be characterized by the corresponding component of *Maxwell's stress tensor* [JUF 95b]:

$$dF = \mu_0 H_t H_n dS \quad [11.3]$$

$$dF/dS = \tau = B_n H_t = \text{tangential surface force} \quad [11.4]$$

B_n is the perpendicular component of the gap flux density;

H_t is the tangential component of the magnetic field in the gap.

The normal component in a permanent magnet motor is created by the magnet and the tangential component via the stator winding.

Based on a sine field distribution and an optimum phase difference equal to zero (optimum performance commutation angle), the tangential final surface force expression becomes:

$$\tau = \tfrac{1}{2}\,\hat{B}_n \hat{H}_t \qquad [11.5]$$

11.1.3. *Permanent magnet motors*

The gap normal flux density component is created by the magnet, its expression is given in [JUF 95a, Chapter 3]:

$$\hat{B}_n = \frac{B_0}{1 + \mu_{dr} \dfrac{\delta_{eq} d_a}{l_a d_\delta}} \qquad [11.6]$$

with:

B_0 = remnant flux density of the PM;

μ_{dr} = differential permeability of the PM;

δ_{eq} = equivalent air gap taking into account the iron MMF drop;

d_δ = average gap diameter;

ℓ_a = PM thickness;

d_a = PM average diameter.

The tangential component of the magnetic field is mainly created by the stator winding. For the air-gap, the following expression can be written:

$$\vec{H} = -\vec{grad}\,\Theta$$

$$H_t = -d\Theta / dy \qquad [11.7]$$

$$\hat{H}_t = \frac{\pi}{\tau_p}\hat{\Theta} \text{ for a sinusoidal MMF distribution} \qquad [11.8]$$

[JUF 99] describes the corresponding expression of this field component:

$$\hat{H}_y = \hat{H}_t = \sqrt{2} J k_w \gamma d_a \left(1 - \frac{\hat{B}_\delta}{B_d k_f}\right) k_{cu} (1+\gamma) \qquad [11.9]$$

γ = ratio slot height/boring diameter.

The surface forces are represented in Figure 11.1 for the following conditions:

d_a = varies from 20 to 200 mm;

$\ell_a = d_a$ (axial length equal to the bore diameter);

$B_o = 0.75$ (Nd Fe B plastic bounded);

$B_d = 1.6$ T.

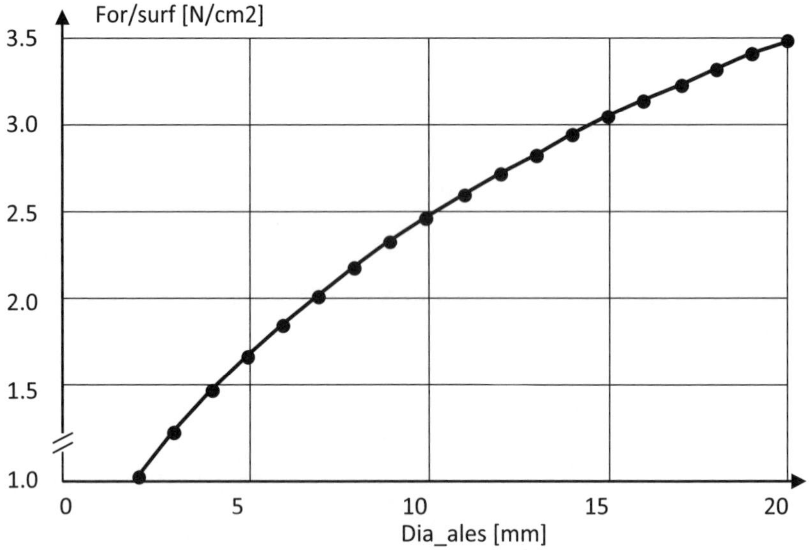

Figure 11.1. *PM motor – specific surface force function of the diameter*

Figure 1.11 shows the surface force evolution as a function of the diameter. This force varies from 1.01 to 3.49 N/cm² for the respective diameters of 20 and 200 mm.

The main associate values are described in Table 11.1.

Diameter	[mm]	**20**	**100**	**200**
τ	[N/cm²]	1.01	2.48	3.49
$\hat{H}_t$	[kA/m]	33.7	82.6	116
J	[A/mm²]	12.73	5.69	4.02
γ	[-]	0.35	0.30	0.26
k_{cu}	[-]	0.34	0.46	0.55
M	[Nm]	0.13	38.9	439
P_{3000}	[kW]	0.04	12.2	138
η	[%]	71.9	94.7	97.3
M/mass	[Nm/kg]	1.47	3.92	5.90
k_{dem}		10.8	2.3	1.4

Table 11.1. *PM motor – performance function of the diameter*

P_{3000} = mechanical power at 3,000 rpm

η = efficiency at P_{3000}

M/mass = mass torque (active mass)

k_{dem} = safety demagnetization coefficient

$$k_{dem} = \hat{\Theta}_{PM} / \hat{\Theta}_{stator} \qquad [11.10]$$

11.1.4. *Induction motor*

For an induction motor, it is also possible to start from the tangential force expression at the rotor surface:

$$\tau = \tfrac{1}{2}\, \hat{B}_n \hat{H}_t \cos\psi \qquad [11.11]$$

The angle ψ corresponds to the shift angle between the magnetization flux and the rotor MMF. The flux density $\hat{B}_n$ is created by the magnetization current component I_m defined by the equivalent electric scheme of Figure 11.2 (see section 4.3.2):

$$\hat{B}_n \approx I_m \approx \hat{\Theta}_m$$

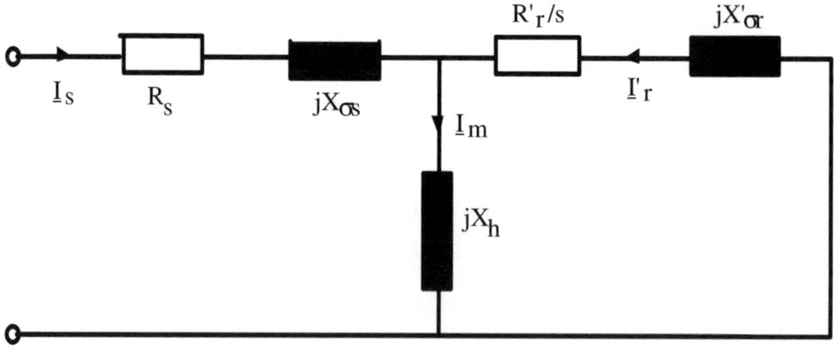

Figure 11.2. *Induction motor equivalent scheme*

The I_m current is part of the I_s current:

$$I_m = \alpha\, I_s \qquad [11.12]$$

according to the motor size:

$$1 > \alpha > 0.25 \qquad [11.13]$$

– for a motor of 1 kW: $I_m \cong 0.5\, I_s$;

– for a motor of 100 kW: $I_m \cong 0.3\, I_s$.

Finally:

$$\hat{\Theta}_s = \hat{\Theta}_t / \alpha \qquad [11.14]$$

$$\hat{\Theta}_m = \frac{\hat{B}_n \delta_{eq}}{\mu_0} \qquad [11.15]$$

δ_{eq} is the equivalent gap taking into account the real gap and the iron effect of the stator and the rotor:

$\delta_{eq} = \delta\, k_c\, k_{sat} = \delta\, k_\delta$;

k_c = Carter' factor = 1.1 to 1.3;

k_{sat} = saturation factor = $(\Theta_{fer} + \Theta_{air}) / \Theta_{air}$ = 1.5 to 2.

In an initial approximation, $\delta_{eq} \simeq 2\delta$.

[JUF 99] describes the equations that determine the surface force.

The characteristic equation is:

$$\tau = \frac{p k_\delta \delta}{\mu_0 \alpha d} \hat{B}_n^2 \cos\psi \qquad [11.16]$$

The important role of the gap flux density is emphasized.

Figure 11.3 presents the evolution of the surface force according to the boring diameter. It should be noted that this is lower than that which is obtained for the magnet synchronous motor. However, for the latter, the demagnetization factor is too small for large diameters, which would require a reduction of the surface force.

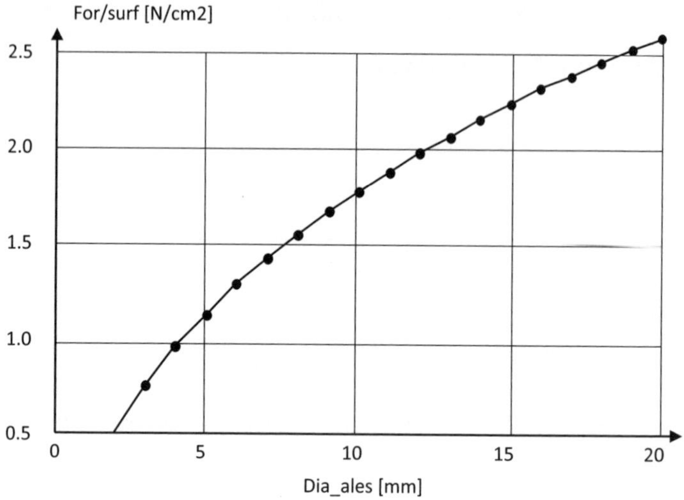

Figure 11.3. *Induction motor – specific surface force according to the diameter*

Table 11.2 describes the main values associated with 3 different diameters.

Diameter	[mm]	**20**	**100**	**200**
τ	[N/cm^2]	.52	1.79	2.60
$\hat{H}_t$	[kA/m]	37.1	50.7	60.6
J	[A/mm^2]	12.7	5.69	4.02
γ	[-]	.35	.30	.26
k_{cu}	[-]	.34	.46	.55
T	[Nm]	.07	28.1	327
P_{1500}	[kW]	0.10	4.42	51.4
η	[%]	12.5	81.4	93.8
M/mass	[Nm/kg]	.43	1.58	2.50
$\hat{B}_\delta$	[T]	.43	.83	.90

Table 11.2. *Induction motor – performances according to the diameter*

P_{3000} = mechanical power at 1,500 rpm, 50 Hz;

B_δ = peak induction in the air gap.

11.1.5. *Comparison*

The suggested methodology enables two approaches to the analysis of the direct drive motors:

– highlighting all the parameters, such as those of current density, quality of the magnets, air-gap, etc. on the performances;

– comparing the various solutions.

In the first case, we can draw the following conclusions:

– for a PM synchronous motor, the parameters determining the performances are the current density, the relative tooth height, the boring diameter, the copper filling factor and the gap flux density created by the PM;

– for an asynchronous motor, the determining values are the reverse of the air-gap, the reverse of the polarity, the square of the current density, the slot height, the cube of the boring diameter, the square of the copper filling factor and the phase shift between the flux and current.

The comparison between the values obtained for the magnet motor as well as for the induction motor makes it possible to draw the following conclusions:

– Figure 11.4 presents the evolution of the efficiency as a function of the diameter for the two types of motors. Clearly, the magnet synchronous motor presents a better efficiency for all the diameters.

– As already reported, the surface force is higher for the synchronous motor than for the induction motor. However, this point must be corrected by the following remark: for large diameters, the coefficient of demagnetization of the magnet motor is too small (1.4 for d = 200 mm). This is not acceptable and would lead to a reduction of the performances.

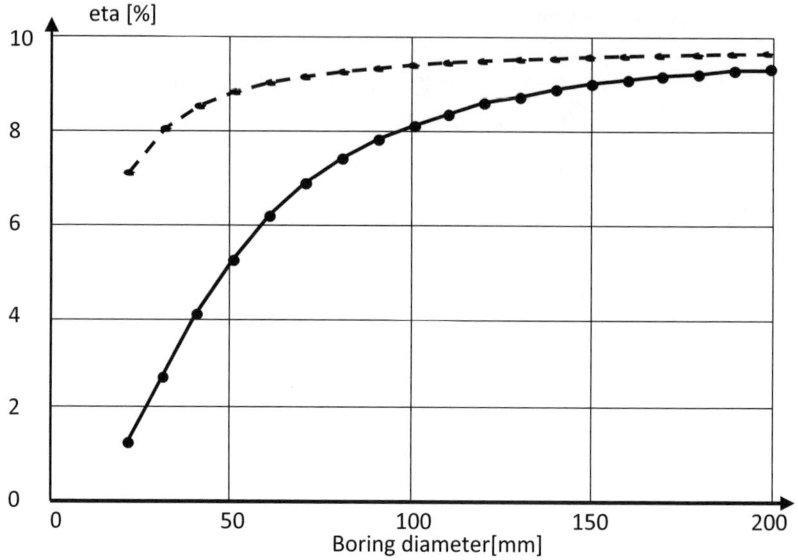

Figure 11.4. *PM and induction motors – efficiency comparison according to the diameter*

11.1.6. *Linear PM motor*

Contrary to the other cases, the boring diameter cannot be the principal reference value. In this case, the pole pitch is considered for this role. The majority of the expressions for the synchronous rotating motor remain valid, by expressing them according to the pole pitch. The principal modifications are as follows:

– for the current density, according to the pole pitch τ_p:

$$J = k_J \sqrt{\tau_p} \quad [11.17]$$

– for the magnetic stator potential (MMF):

$$\Theta = \sqrt{2/\pi} k_w \gamma' J \tau_p^2 \left(1 - \frac{\hat{B}_\delta}{B_d k_f}\right) k_{cu} \qquad [11.18]$$

with γ' = relative slot height,

$$\gamma' = h_N/\tau_p = 1/(1 + 10\,\tau_p) \qquad [11.19]$$

– the resulting force is calculated in the following conditions:

$\ell_a = \tau_p$, 2 poles

– frequency $f = 50$ Hz, with for the nominal speed

$$\upsilon_N = 2 f \tau_p \qquad [11.20]$$

– the chosen air gap is greater than that of a rotating motor.

Figure 11.5 shows the evolution of the surface force according to the pole pitch. In a surprising way, this decreases after having passed by a maximum. This is due to the limitation of the coefficient of demagnetization of the magnets, fixed at 1.5. This limitation appears for a pole pitch of 110 mm. A high value of τ for some small pole pitch values is observed. Slots that are higher and easier to wind make this possible.

However, this phenomenon is accompanied by a significant fall in the current density, which involves an improvement of the efficiency. This is particularly highlighted by Figure 11.6.

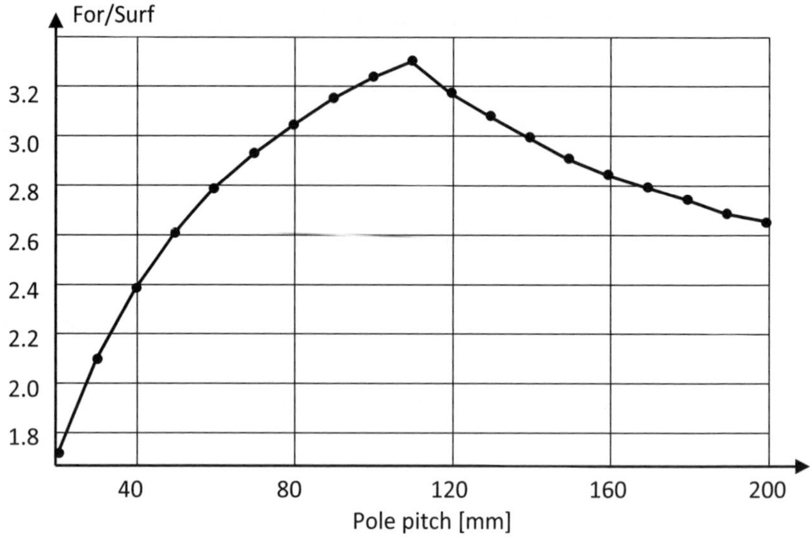

Figure 11.5. *PM linear motor – specific surface force function of the pole pitch*

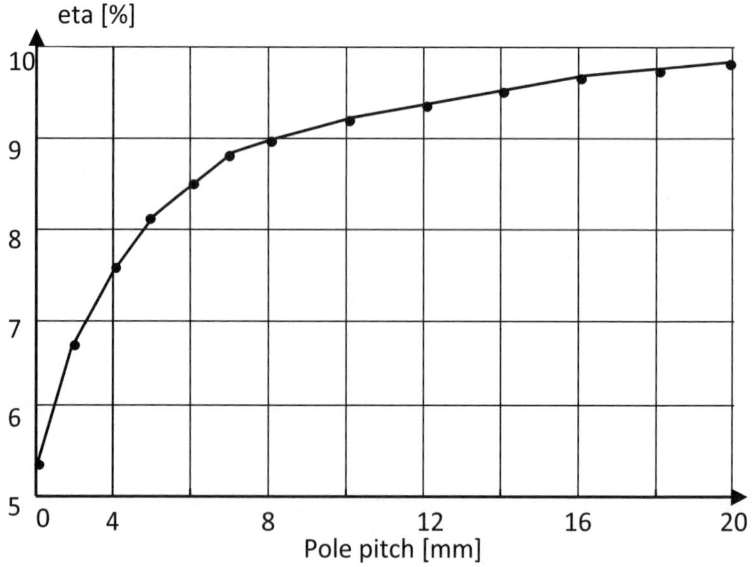

Figure 11.6. *PM linear motor – efficiency function of the pole pitch*

Pole pitch [mm]		20	100	200
τ	[N/cm^2]	1.75	3.19	2.63
$\hat{H}_t$	[A/m]	58.45	106.2	87.8
J	[A/mm^2]	12.7	5.7	2.9
γ	[-]	.83	.50	.33
k_{cu}	[-]	.34	.46	.55
F	[N]	14	639	2107
P	[kW]	0.028	6.39	42.15
η	[%]	53.1	92.3	97.5
F/masse	[N/kg]	98.2	51.0	26.2
k_{dem}	[-]	8	1.6	1.5

Table 11.3. *PM linear motor – performances function of the pole pitch*

Table 11.3 shows the main results for 3 pole pitches, 20, 100 and 200 mm.

11.1.7. *Conclusions*

The method presented is well adapted for a comparative analysis of the solutions and the design of direct drives.

Though practical and effective, this kind of method nevertheless introduces a certain number of coefficients resulting from the experiment (slot height, air-gap, filling factor) or requiring a process of iteration (the coefficient of demagnetization, the factor α, the coefficient of saturation).

Its advantages make it interesting in the search for optimal performances.

11.2. Motor with external rotor

11.2.1. *Specifications*

Traditionally, a rotating motor is characterized by an external stator, associated with a housing structure supporting bearings and an internal rotor with its shaft. It is possible to reverse the position on these two components if the rotor is a direct support trained body that then wraps around the rotor. This is, for example, the case for most car radiator cooling fans: the rotor is the support for the fan fins (Figure 11.7). Another possible application is the wheel motor for which the external rotor is the rim and tire support (Figure 11.8).

If the general structure of an external rotor motor is more complex in terms of bearings and protection, an important advantage is inherent. The torque by mass unit is significantly higher than for the traditional solution.

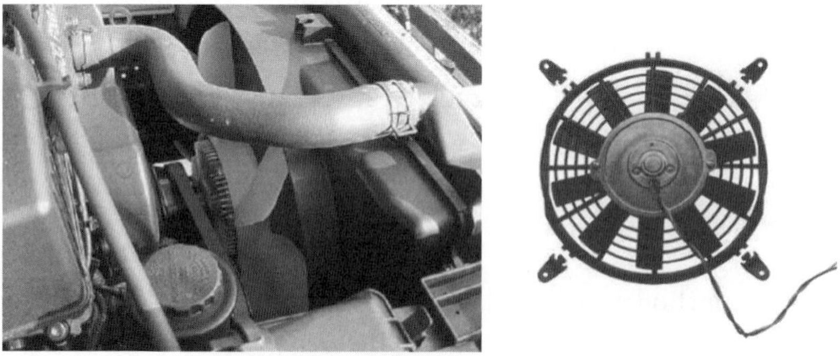

Figure 11.7. *External rotor motor for fans*

11.2.2. *Torque*

Referring to the scaling laws, it has been shown that in the case of a permanent magnet, motor torque is proportional to the power of 2.5 (relation [5.38]) of the bore diameter.

For a typical internal rotor motor, the bore diameter equals an external diameter of half the housing thickness, stator yoke thickness and slot height.

$$d_a = d_{ext} - 2(e_{car} + e_{cul} + h_{enc}) \qquad [11.21]$$

Typical values are:

$$e_{car} = 0.03 \div 0.07 d_a = \text{housing thickness}$$

$$e_{cul} = 0.05 \div 0.1 d_a = \text{yoke thickness}$$

$$h_{enc} = 0.07 \div 0.2 d_a = \text{slot height} \qquad [11.22]$$

$$d_a = d_{ext} - 0.3 \div 0.74 d_a$$

$$d_a = 0.77 \div 0.575 d_{ext}$$

In return for an external rotor motor, the bore diameter is the external diameter reduced by the double rotor yoke thickness and PM thickness:

$$d_a = d_{ext} - 2(e_{cul} + e_{ai})$$

$$e_{cul} = 0.05 \div 0.1 d_a = \text{rotor yoke thickness}$$

$$e_{ai} = 0.015 \div 0.025 d_a = \text{PM thickness} \qquad [11.23]$$

$$d_a = d_{ext} - 0.13 \div 0.25 d_a$$
$$d_a = 0.885 \div 0.8 d_{ext}$$

Referring to the extreme cases, torque ratio between the external rotor and internal rotor for the same diameter is:

$$M_{ext} / M_{\text{int}} = 1.42 \div 2.28 \qquad [11.24]$$

The external rotor variant is therefore much more interesting in terms of torque per volume. In addition, a relatively large volume becomes available in the stator (diameter d_{int}), particularly for slow motors (Figure 11.8). Figure 11.9 shows a possible use for this free area, namely the insertion of the damper system.

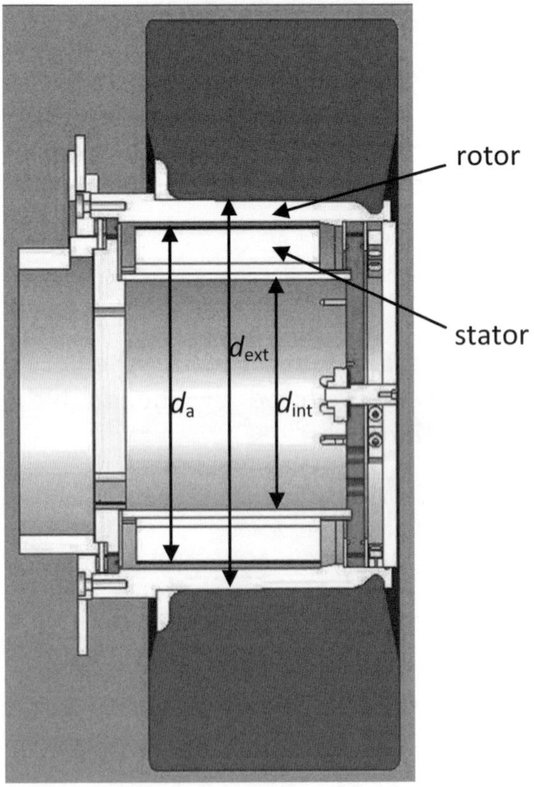

Figure 11.8. *Wheel-motor, with external rim rotor – diameter definition*

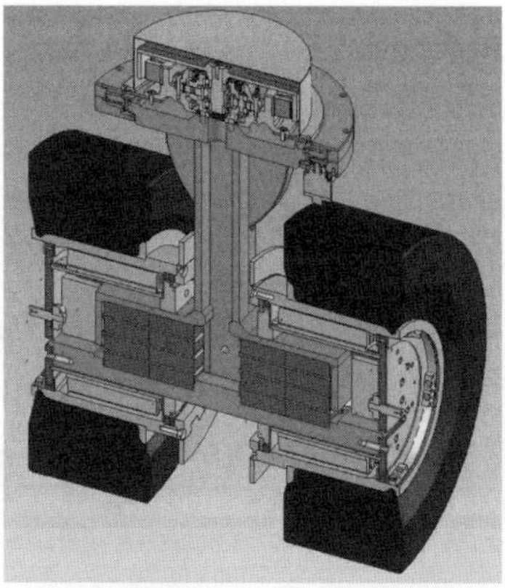

Figure 11.9. *Axis with wheel-motors, with shock absorbers in the stator*

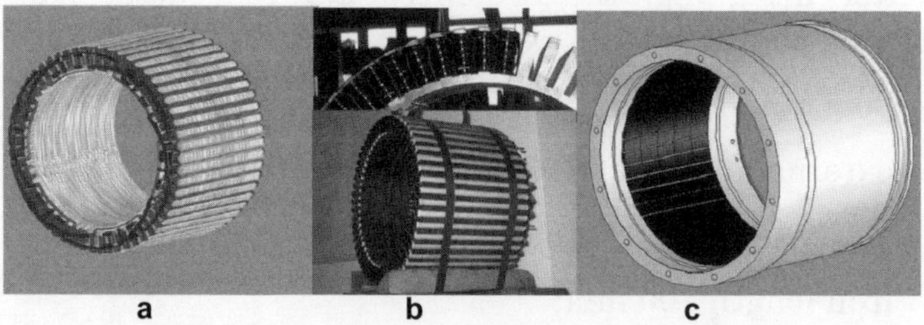

Figure 11.10. *a) External rotor motor; b) stator; c) rim rotor with PM*

Figure 11.10 shows the details of the stator and rotor for this kind of wheel-motor.

11.3. Example

11.3.1. *Specifications*

An automatic vehicle handling load of 40 tonnes (containers) is characterized by the following performance:

– 24 inch rims, with an external diameter of 955 mm;

– rated speed 18 km/h;

– reference torque 1,840 Nm;

– each wheel must be equipped with its own electric drive, based on a possible deflection of 270 degrees.

Two solutions are possible:

– a wheel-motor direct drive, with external rotor and permanent magnets;

– a high speed (3,000 to 6,000 rpm) motor, associated with a two-level gearbox.

11.3.2. *External wheel-motor*

After designing, the brushless DC PM motor has the following characteristics:

– pole number 48;

– external diameter 462 mm;

– active length 108 mm;

– iron length 100 mm;

– bore diameter 440 mm;

– surface force [11.5 to 11.9] 5.5 N/cm^2;

– active mass 51 kg;

– efficiency 84.9%.

11.3.3. *Classic motor with gearbox*

A 4,500 rpm speed is chosen for the motor. After sizing, the corresponding PM motor, and brushless DC type motor, has the following characteristics:

– pole number = 4;

– rated torque = 45.4 Nm;

– rated speed = 4,500 rpm;

– external diameter = 180 mm;

– external length = 167 mm;

– boring diameter = 105 mm;

– active length = 100 mm;

– surface force (Figure 11.1) = 2.55 N/cm^2;

– active mass = 16 kg;

– motor efficiency = 97.3%;

– transmission = 2 stage gearbox of ratio 7.5 and 6;

– global efficiency = 86%.

Figure 11.11 shows the structure of such a solution with gearbox. Integration in the wheel requires us to shift the motor laterally under the floor of the vehicle, which complicates its access.

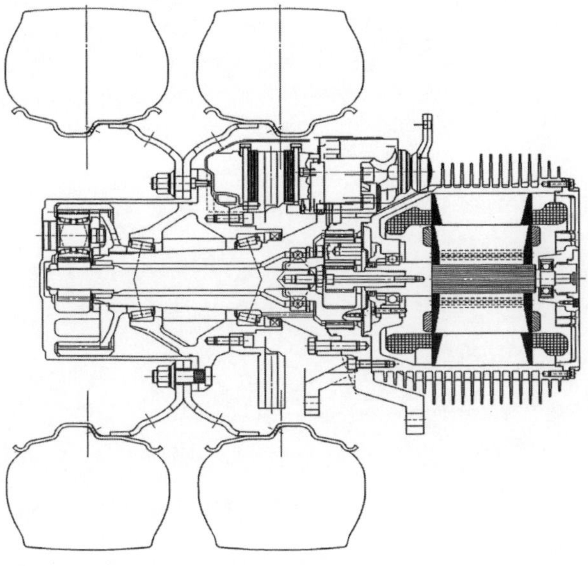

Figure 11.11. *Classic fast motor with two-stage gearbox*

11.3.4. *Choice*

The choice between these two solutions will focus on economic criteria, integrating the components cost and maintenance aspects. In particular, the total mass – motor on one hand and motor + gearbox on the other hand – are very close. However, maintenance is of more interest for the solution without a gearbox.

Other criteria such as cooling may also influence the choice.

Chapter 12

Integrated Drives

12.1. Principle

In a traditional design, mechanical functions (gearbox, screw, belt, brake, etc.) and electromechanical functions are grouped with the driven body, while electrical power and control functions are remote and grouped in an *ad hoc* cabinet.

Possibilities for integration of power and control electronics today enable us to implement these components within the motor itself. It then becomes an *integrated electric drive*, sometimes called a *smart motor* [JUF 92]. This approach mainly concerns the brushless DC motor.

The main benefits are an important volume and cabling saving. Moreover, the distance between power and motor being both short and invariant, the impedance line is constant and no electrical parameter adaptation is necessary.

The main drawback is the merging of two losses, i.e. heat sources: motor and electronic commutation. This limits the possible power values to the order of a few kW. Beyond that, forced air or water circulation cooling becomes necessary.

12.2. Realization

12.2.1. *Motor and electronics*

The integral functions in a motor that form an integrated electric drive are (Figure 12.1):

– motor;

– power electronic;

– control electronic;

– communication with the centralized control system (multi-axis control);

– the microprocessor integrating various settings;

– sensors.

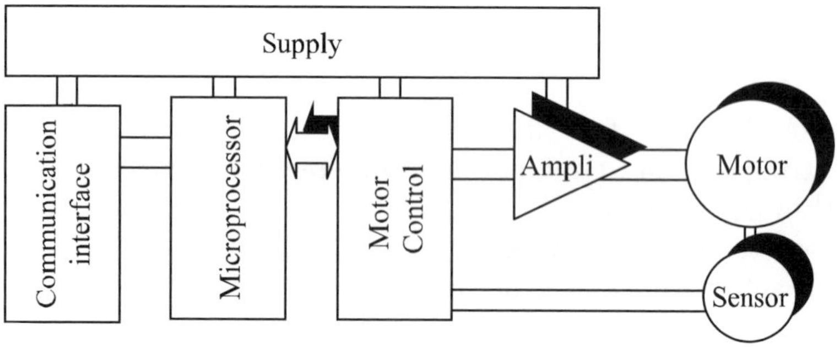

Figure 12.1. *All functions are contained within an integrated electric drive*

Figure 12.2 shows the picture of this kind of motor with respective electronics (clear) and motor (dark) housing. The grooves used to increase the convection of these elements are visible.

Figure 12.2. *Integrated drive with respective convection grooves for electronics and motor*

Figure 12.3 shows the cross-section view of this assembly. The only necessary connections are a DC power supply and cable for information transmission in both directions in the form of a field bus.

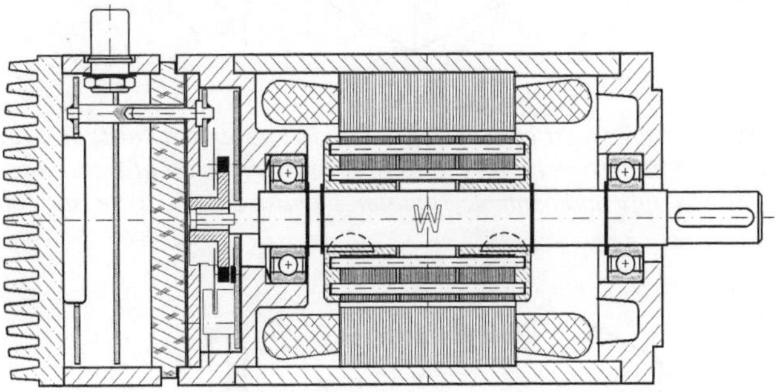

Figure 12.3. *Integrated drive with respective housings for electronics and motor. Between them is a high-resolution sensor*

12.2.2. *System*

The comparison between a system of motors supplied from a centralized control cabinet and integrated electric drives can be highlighted by an example of application for an industrial robot. Figure 12.4 shows all connections from the central cabinet to each motor with three-phase cabling and network cable routing information (position, speed, temperature) sensors.

Figure 12.5 shows the same system using integrated drives. Just one cable with two conductors routes DC power to the motors, with branches in parallel. A field bus-type cable can route orders to a central computer and inform the various axes of actual specifications.

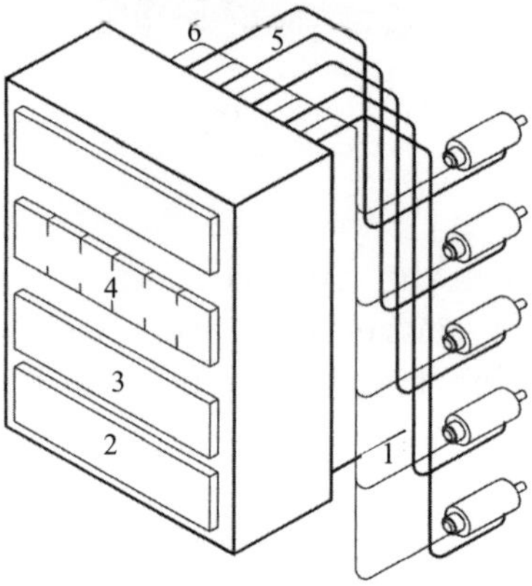

Figure 12.4. *Power and control of a 5 axis robot to a central cabinet and corresponding cabling: 1) AC supply; 2) transformer-rectifier; 3) axis control; 4) motor supply and control; 5) motor 3-phase cables; 6) sensor cables*

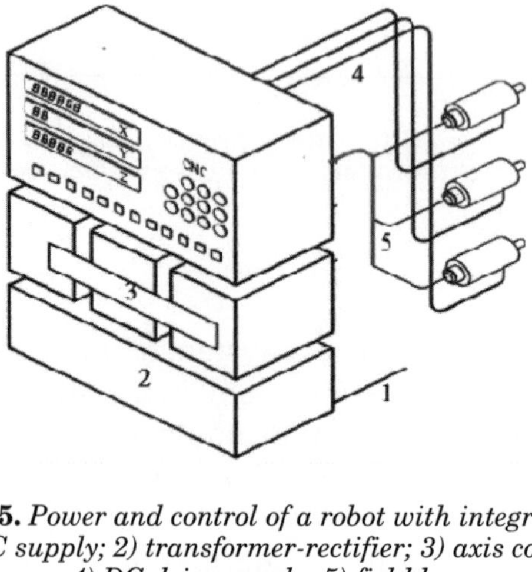

Figure 12.5. *Power and control of a robot with integrated drives: 1) AC supply; 2) transformer-rectifier; 3) axis control; 4) DC drive supply; 5) field bus*

12.2.3. *Transmission integration*

For large scale applications, the transmission system can also be integrated, allowing us to remove one bearing and reduce the drive length. Figure 12.6 shows a possible modular principle. Figure 12.7 shows this type of motor cross-section with, successive from left to right, electronics, sensor, motor and two stages of transmission modules.

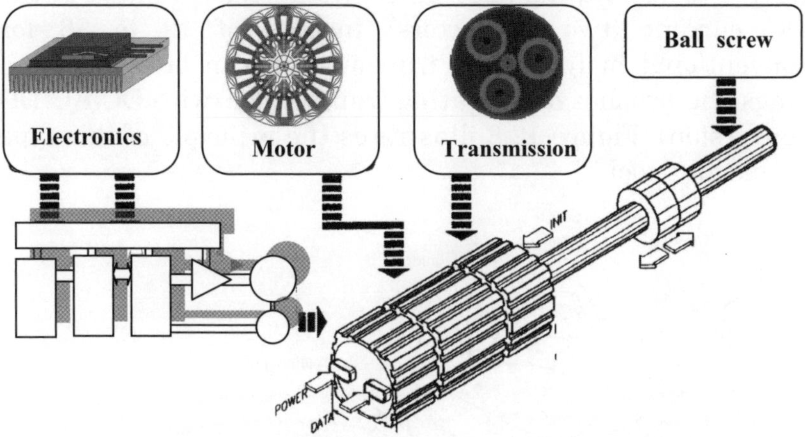

Figure 12.6. *Integrated modular design of an electric drive, including transmission*

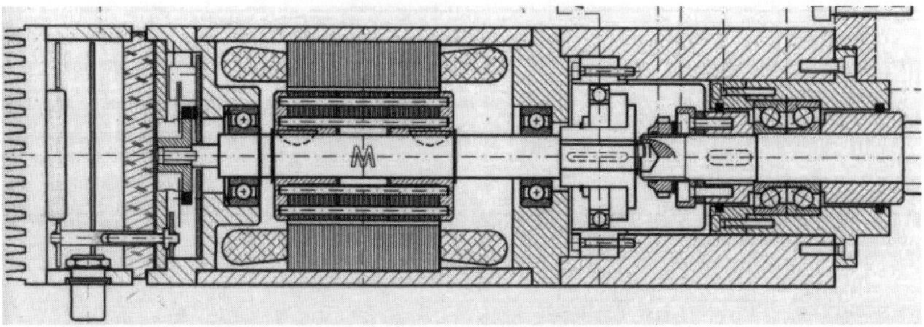

Figure 12.7. *Cross-section of an integrated drive including transmission*

12.2.4. *Applications*

Applications for special machine tools are particularly interesting. For example, a finishing machine (grinding, polishing) has many very important advantages of using integrated drives. In addition to tool axis drive, the integrated drive's use of machined piece pin support on a rotary table represents a decisive advantage. The table can have 4 to 12 rotating axes. Energy transfer is limited to the DC current (two conductors) instead of 12 to 36 for conventional motors. This transmission can be made with rings and brushes or a rotating transformer (with DC-AC-DC conversion). Figure 12.8 illustrates the principle of this type of machine tool.

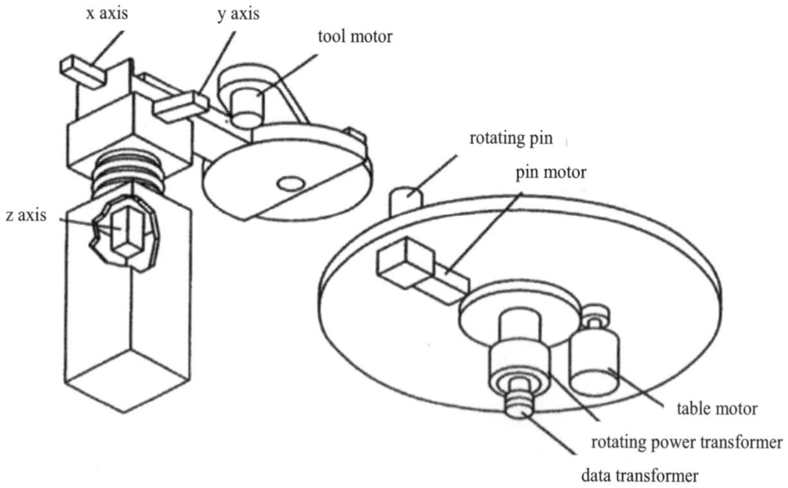

Figure 12.8. *Finishing machine-tool*

Symbols

a	[m]	length
a	[m/s^2]	acceleration
b	[m]	length
B	[T]	magnetic flux density
c	[m]	length
c	[J/Km^3]	specific heat per volume
C	[F]	condenser capacitance
d	[m]	diameter
e	[m]	thickness
f	[Hz]	frequency
F	[N]	force
g	[m/s^2]	earth gravitation
h	[m]	height
H	[A/m]	magnetic field
i,I	[A]	current

j,J	[A/m^2]	current density
J	[kgm2]	inertia
k	[rad/m]	rotating-linear transmission ratio
k		coefficient
l	[m]	length
L	[H]	inductance
m	[kg]	mass
M	[Nm]	moment, torque
Np	[-]	number of steps per revolution
N	[-]	number of turns in series
N	[t/min]	rotation speed
p,P	[W]	power
p_r	[-]	relative precision
p	[-]	pole pair number
r	[-]	rotating transmission ratio
r	[-]	scaling ratio
R	[Ω]	resistance
r,R	[m]	radius
s	[-]	slip
S	[VA]	apparent power
S	[m^2]	surface
t	[s]	time
T	[s]	period

T	[K]	temperature, heating
u,U	[V]	voltage
$ü$	[-]	transformer ratio
v	[m/s]	speed
V	[m^3]	volume
W	[J]	energy
x,y,z	[-]	coordinate
X	[Ω]	reactance
Z	[Ω]	impedance
α	[rad]	angle
α	[-]	coefficient
α	[W/Km^2]	convection coefficient
γ	[-]	relative slot height
δ	[m]	air gap
ε	[rad/s2]	angular acceleration
ε	[rad]	angle
φ	[rad]	angle
ψ	[rad]	angle
η	[-]	efficiency
λ	[W/Km]	thermal conductivity
Λ	[Vs/A]	permeance
μ	[Vs/Am]	permeability
ρ	[Ωm]	resistivity

χ	[Ns/m]	linear viscous friction coefficient
ξ	[Nms]	rotating viscous friction coefficient
τ	[N/m^2]	tangential surface force
Θ	[A]	MM force
ω	[rad/s]	pulsation
Ω	[rad/s]	angular rotation speed

Indices

a	active
a	boring
ai	relating to permanent magnet
amb	ambient
at	with autotransformer
c	on a cycle
c	convection
cc	short-circuited
co	conductor
cu	copper
kin	kinetic
d	at start-up
d	differential
e	driven body
e	excitation

el	electrical
enc	slot
ext	external
f	braking
h	harmonic
h	mutual
i	induced
int	internal
J, Copper	copper losses
k	rotating-linear transmission
K	maximum or minimum
l	line
L	inductance
lim	limit
m	time out
max	maximum
mec	mechanical
min	minimum
M	torque
n	normal
N	nominal, rated
osc	oscillatory
p	step related

ph	phase
r	rotating-rotating transmission
r	load torque
r	relative
r	rotor
r	network
rel	reluctance
s	stator
s	synchronous
t	related to the transmission
t	tangential
tot	total
ut	useful
x	x direction
0	at origin
1	time period 1
2	time period 2
δ	air gap
Δ	triangle coupling
λ	star coupling
σ	leakage

Bibliography

[CAR 93] CARDOLETTI L., Commande et réglage de moteurs synchrones auto-commutés par des capteurs indirects de position, Thesis, EPFL 1118, 1993.

[FAV 04] FAVRE E., TEPPAN W., "Capteurs de courant : A chacun son argument", *LEM Mesures 763*, March, 2004, available online at: http://www.mesures.com/archives/049_O54_SOL.pdf.

[JUF 89] JUFER M., *Entraînements électriques*, 3rd edition, EPFL course, EPFL, Lausanne, 1989.

[JUF 92] JUFER M., "Smart motor technology - Advantages and performance comparison", *Proceedings 21st Annual Symposium on Incremental Motion Control Systems and Devices* (IMCSD), p. 44-52, Champaign, IL., USA, June 1992.

[JUF 95a] JUFER M., *Electromécanique,* Traité d'électricité, vol. 9, 3rd edition, Presses Polytechniques et Universitaires Romandes, Lausanne, 1995.

[JUF 95b] JUFER M., "Indirect Sensors for Electric Drives", *Proceedings 6th European Conference on Power Electronics and Applications* (EPE'95), vol. 1, p. 1 836-1 841, Seville, Spain, 1-19 September 1995.

[JUF 96] JUFER M., "Design and Losses - Scaling Law Approach", *The Nordic Research Symposium Energy Efficient Electric Motors and Drives*, p. 21-25, Skagen, Denmark, August 1996.

[JUF 99] JUFER M., "Performances limites des entraînements directs", *Proceedings CEMD'99*, p. 15-21, Cachan, February 1999.

[OSS 88] OSSENI R., Modélisation et autocommutation de moteurs synchrones, Thesis, EPFL 767, 1988.

Index